T0252517

Grid-connected Solar Electric

Solar electricity – or photovoltaics (PV) – is the world's fastest growing energy technology. It can be used on a wide variety of scales, from single dwellings to utility-scale solar farms providing power for whole communities. It can be integrated into existing electricity grids with relative simplicity, meaning that in times of low solar energy, users can continue to draw power from the grid, while power can be fed or sold back into the grid at a profit when their electricity generation exceeds the amount they are using.

The falling price of the equipment combined with various incentive schemes around the world has made PV into a lucrative low carbon investment, and as such demand has never been higher for the technology, and for people with the expertise to design and install systems.

This Expert handbook provides a clear introduction to solar radiation, before proceeding to cover:

- electrical basics and PV cells and modules
- inverters
- design of grid-connected PV systems
- system installation and commissioning
- maintenance and troubleshooting
- health and safety
- economics and marketing.

Highly illustrated in full colour throughout, this is the ideal guide for electricians, builders and architects, housing and property developers, home owners and DIY enthusiasts, and anyone who needs a clear introduction to grid-connected solar electric technology.

Geoff Stapleton has been instrumental in developing training and capacity building both in Australia and overseas, particularly in Ghana, Sri Lanka, Malaysia and China. He set up Global Sustainable Energy Solutions Pty Ltd as a renewable energy training and consultancy business in 1998 and is a part-time lecturer at University of New South Wales, Australia.

Susan Neill has worked in the renewable energy industry for over 25 years. She is now director of training and engineering for Global Sustainable Energy Solutions and is a guest lecturer at UNSW, Australia.

Grid-connected Solar Electric Systems

The Earthscan Expert Handbook for Planning, Design and Installation

Geoff Stapleton and Susan Neill

SERIES EDITOR:
FRANK JACKSON

Routledge
Taylor & Francis Group

LONDON AND NEW YORK

earthscan
from Routledge

First published by Earthscan in the UK and USA in 2012

For a full list of publications please contact:
Earthscan
2 Park Square, Milton Park, Abingdon, Oxon OX14 4RN
605 Third Avenue, New York, NY 10017

First issued in paperback 2021

Earthscan is an imprint of the Taylor & Francis Group, an informa business

© Global Sustainable Energy Solutions Pty Ltd, 2012. Published by Taylor & Francis

The right of Geoff Stapleton and Susan Neill to be identified as the authors of this work has been asserted by them in accordance with sections 77 and 78 of the Copyright, Designs and Patents Act 1988.

All rights reserved. No part of this book may be reprinted or reproduced or utilized in any form or by any electronic, mechanical or other means, now known or hereafter invented, including photocopying and recording, or in any information storage or retrieval system, without permission in writing from the publishers.

Notices
Practitioners and researchers must always rely on their own experience and knowledge in evaluating and using any information, methods, compounds, or experiments described herein. In using such information or methods they should be mindful of their own safety and the safety of others, including parties for whom they have a professional responsibility.

Product or corporate names may be trademarks or registered trademarks, and are used only for identification and explanation without intent to infringe.

While the authors and the publishers believe that the information and guidance given in this work are correct, all parties must rely upon their own skill and judgement when making use of them – it is not meant to be a replacement for manufacturer's instructions and legal technical codes. Neither the authors nor the publisher assume any liability for any loss or damage caused by any error or omission in the work. Any and all such liability is disclaimed.

British Library Cataloguing in Publication Data
A catalogue record for this book is available from the British Library

Library of Congress Cataloging in Publication Data
Stapleton, Geoff.
Grid-connected solar electric systems the Earthscan expert handbook for planning, design, and installation/Geoff Stapleton and Susan Neill.
p. cm.
Includes bibliographical references and index.
1. Building-integrated photovoltaic systems–Installation. 2. Electric power distribution. I. Neill, Susan. II. Title.
TK1087.S83 2011
621.31'244–dc22
2011008240

Typeset in Sabon
by Domex e-Data, India

ISBN 13: 978-0-367-78748-6 (pbk)
ISBN 13: 978-1-84971-344-3 (hbk)

Contents

Illustrations

Figures

Tables

Boxes

Preface and Acknowledgements

The worldwide market for grid-connected solar electric systems has increased from 1.55 gigawatts (GW) installed in 2006 to 11.86GW in 2010. This 2010 figure represents an increase of 665 per cent over the 2006 figure.

It is to be expected that the general public as well as tradesmen, technicians and other professionals will need information about all aspects of grid-connected solar as they see these systems installed in their suburbs and on larger roof spaces; they will also wish to know how it affects their lives.

In the absence of basic technology and installation information, how and why grid-connected solar electric systems work and the value they can represent for the electricity grid may be misrepresented. As can be seen by the increase in the grid-connected solar electric market, the technology and production of solar modules and enabling products (e.g. inverters, mounting structures etc.) are now mature; product demand has been increasing from year to year with healthy forward projections, so many more manufacturers have moved into this technology market, driving prices down; governments around the world have introduced various economic drivers for renewable energy, with those affecting solar electric systems being typically subsidies, feed-in tariffs and redeemable renewable energy credits.

Global Sustainable Energy Solutions was pleased to be approached by Earthscan to write this book now, because the market demand for the product clearly demonstrates a need for information at all levels.

This book is suitable for anyone wanting to learn about grid-connected solar electric systems starting with the explanation of solar radiation, its origin, solar modules, solar electric systems, system composition, system installation, through to the economics of these systems.

Thanks are due to a number of people who have contributed time and/or information during the evolution of this book: Caitlin Trethewy, who has worked from day one on this publication collecting information, researching, writing content and progressively editing the chapters to completion; the staff at Global Sustainable Energy Solutions Pty Ltd who have contributed to the book's development; Pamela Silva for her contribution while in Australia from the US; Anthony Allen for his technical drawing skills; all image providers, whom we trust are correctly acknowledged; the companies and individuals who have provided us with case study information and photos as follows: Blair Reynolds, BMC Solar (www.bmcsolar.com), Briana Green and Green Solar Group (www.greensolargroup.com.au), Frank Jackson, Paul Barwell, Tony J. Almond and Planet Energy Solutions (www.planetenergy.co.uk).

Acronyms and Abbreviations

AC	alternating current
ANU	Australian National University
ASCE	American Society of Civil Engineers
BIPV	building-integrated photovoltaics
BCSC	buried contact solar cells
BoS	balance of system
CCC	current-carrying capacity
CdTe	cadmium telluride
CIGS	copper-indium-gallium-diselenide
CSI	California Solar Initiative
CSP	concentrated solar power
CVD	chemical vapour deposition
DC	direct current
DNO	distribution network operator
ELV	extra low voltage
ESTI	European Solar Test Installation
FiT	feed-in tariff
GHG	greenhouse gas
GSES	Global Sustainable Energy Solutions
HSE	health, safety and environment
HIT	heterojunction with intrinsic thin layer
IEC	International Electrotechnical Commission
IEEE	Institute for Electrical and Electronics Engineers
I_{mp}	current at maximum power point
IP	ingress protection
IREC	Interstate Renewable Energy Council
I_{sc}	short-circuit current
LED	light-emitting diode
LV	low voltage
MCB	miniature circuit breaker
MOV	metal oxide varistor
MPPT	maximum power point tracker
NABCEP	North American Board of Certified Energy Practitioners
NEC	National Electric Code
NEG	net excess generation
NEMA	National Electrical Manufacturers Association
NOCT	nominal operating cell temperature
NREL	National Renewable Energy Laboratory
$P_{max}/P_{mp}/MPP$	maximum power point
PSH	peak sun hours
PV	photovoltaic
PVGIS	Photovoltaic Geographical Information System
RECs	renewable energy certificates
RHI	Renewable Heat Incentive

ROCs	Renewable Obligation Certificates
RPS	renewable portfolio standard
SDI	single core, double insulated (cable)
SRECs	Solar Renewable Energy Credits
STC	standard test conditions
TRECs	Tradable Renewable Energy Certificates
UEDS	utility external disconnect switch
UL	Underwriters' Laboratory
V_{mp}	voltage at maximum power point
V_{oc}	open-circuit voltage
Wh	watt-hour
Wp	watt-peak

1

Introduction to Solar Power

Almost every aspect of modern life is dependent on electricity. As technology improves, almost everything we use, from cooking and cleaning appliances to sports and entertainment equipment, has been updated to include electronic circuitry. However, society's dependence on electricity is increasingly problematic: as conventional fossil fuel reserves diminish and the danger of climate change materializes, the need for alternative energy sources is unprecedented. Solar power, that is electricity generated using energy from the sun, is an attractive way to offset our reliance on electricity generated by burning fossil fuels.

This book explores how solar energy is converted into electricity that can be used to power our homes and cities. It begins with a description of the solar resource, which is the sun's energy hitting the Earth's surface, and goes on to describe ways in which this resource can be quantified and captured by solar power systems. It explores different technologies used to convert sunlight into electricity and details how a grid-connected photovoltaic system is designed, installed and commissioned at the residential level.

Why solar power?

Since the industrial revolution, developed nations have been burning fossil fuels in ever increasing quantities. However, in the past decade two major problems with this have come to light. The first relates both to the scarcity of fossil fuel resources and the notion of energy security. The oil reserves that the world currently depends on are accepted as being a limited resource, and it is acknowledged that demand for oil will only increase as countries such as India and China increase their levels of industrialization and burgeoning middle classes demand more power. This growth will create an enormous demand on finite resources. Moreover, these resources are generally found in areas characterized by conflict, such as the Middle East, and prices can be highly variable. Interest in renewable energy technologies like solar power is driven by the need to meet these increasing needs and a desire for independence from ever increasing oil prices. The beauty of solar power is that it has very few running costs: once the capital investment has been made the energy is effectively free, while fossil fuels must be purchased continuously and all indications are that prices will continue to rise.

The second problem is that of human-induced climate change (also known as global warming), the widely accepted theory that burning fossil fuels releases greenhouse gases (GHGs) such as carbon dioxide and methane into the

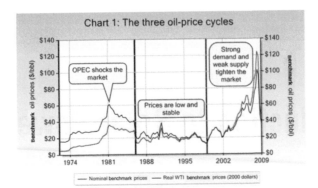

Figure 1.1 The cost of crude oil has been increasing over the last 30 years and is expected to continue increasing indefinitely

Source: Chris Laughton, data from IEA

atmosphere, causing the planet's climate to change and alter weather patterns. Scientists predict more extreme weather events such as floods and droughts will occur as a result of climate change and that the average global temperature will rise. Naturally this affects the plants and animals that inhabit the Earth. Many species are predicted to suffer extinction if urgent action is not undertaken. Human-induced climate change has been acknowledged by all levels of government in developed and developing countries as needing urgent international action to mitigate its effects. This action will include measures to reduce the use of fossil fuels, so reducing the release of GHGs. A major contribution to the success of this action will be substituting fossil fuels with cleaner, low-emission technologies. Renewable energy technologies that make use of wind and solar power will therefore play an important role in reducing global GHG emissions and are likely to represent a large portion of the world's energy production in the future.

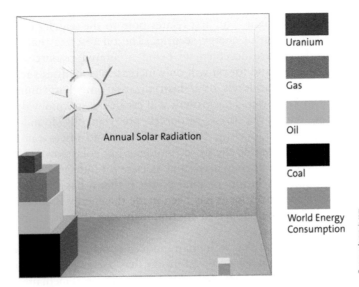

Figure 1.2 The Earth receives more than enough solar radiation each year to meet its current energy consumption 1000 times

Source: www.solarpraxis.com

What is a solar electric system?

Solar energy is harnessed by using solar electric systems, also known as photovoltaic (PV) systems. The word photovoltaic is derived from the Latin words *photo* (light) and *voltaic* (energy). PV devices capture the energy in sunlight and convert it into electricity – that is, they use light energy. Solar electric systems should not be confused with solar thermal systems which use the sun's energy to heat a substance (typically water). Electric and thermal systems are very different in appearance and operation, as Figure 1.3 shows. There are several different types of solar electric systems, discussed below. This book focuses on solar electric systems that feed electricity back into the grid (grid-connected solar electric systems).

Off-grid systems

An off-grid or stand-alone solar electric system is designed to replace or supplement conventional mains power supply. They are most commonly used in rural or remote areas where mains power is not available due to the high cost of grid extension. Off-grid systems use solar power to charge batteries that store the power until it is needed for use. The battery power is then used to operate appliances (e.g. lighting, pumps, refrigeration etc.) either directly from the batteries as low-voltage equipment (e.g. 12V DC lights) or from a power inverter, which is connected to the batteries and converts the battery voltage to mains equivalent for use in regular AC equipment (e.g. computer, television, radio etc.).

Off-grid systems can range in size from a single PV module, battery and controller to very expensive large systems incorporating sophisticated control equipment and large back-up generating sets. Further reading about off-grid systems can be found in Chapter 15.

Figure 1.3 An electrician installing a solar system that includes both a solar hot water system (left) and a solar electric system (right)

Source: Global Sustainable Energy Solutions

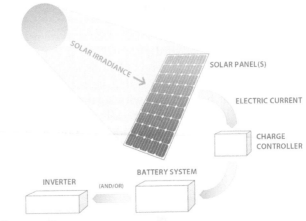

Figure 1.4 Layout of an off-grid system

Source: Global Sustainable Energy Solutions

Figure 1.5 Small off-grid solar electric systems are often the most convenient and cost-effective option for those living in rural areas of the developing world such as the West Sahara, where this system is located

Source: Global Sustainable Energy Solutions

Figure 1.6 Batteries in a stand-alone solar electric system

Source: Global Sustainable Energy Solutions

Grid-connected systems

Grid-connected or utility-interactive solar electric systems are the focus of this book. Unlike off-grid systems they are not designed to be a substitute for grid power. Grid-connected systems are normally found in urban areas that have readily available mains supply and instead of storing the electricity generated by the PV system in batteries, the power is fed back into the grid. In this way the grid acts as a kind of storage medium and when power is needed in the building it can be imported from the grid. One of the key benefits of this is that the system does not have to supply enough electricity to cover the property's power demand as in an off-grid system. The property can be powered by the PV system, the electricity grid or a combination of the two, meaning that the system can be as small or large as the owner desires. Excess power generated by the PV system will be exported to the power grid and in many areas the system owner is paid for the exported power.

The major components of a grid-connected PV system include the PV array, inverter and the metering system. In addition to these major components are the necessary cables, combiner boxes, protection devices, switches, lightning protection and signage.

Figure 1.7 Grid-connected PV systems are becoming more common in urban areas

Source: Green Solar Group

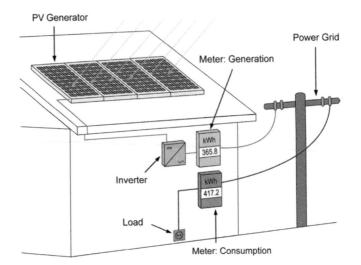

Figure 1.8 Typical small grid-connected PV system (as set up for gross metering). A grid-connected PV system uses the sunlight that strikes the PV array to generate DC power. This DC power is converted to AC power via the inverter and the AC power is supplied (or fed) into the electricity grid via the meter

Source: Global Sustainable Energy Solutions

Central grid-connected PV systems

The existing electricity system typically consists of central power stations using a variety of fuel sources such as coal, gas, water (hydro) or diesel that provide power to end-users via transmission lines and a distribution system. The power stations directly connect to the transmission lines and the power produced by the power stations is consumed by end-users at their actual location in factories, businesses and homes.

A central grid-connected PV system operates in the same way. A large PV array is directly connected to the transmission lines. Grid-connected central PV systems can be as small as 50kWp, while systems as large as 60MWp have been installed in recent years in Europe. There are also companies planning grid-connected PV power plants over 1GWp, with many plants to be completed in the near future.

Figure 1.9 Alicante, Spain, 5.6MWp central PV system

Source: SMA Solar Technology AG

Distributed grid-connected PV systems

As the name suggests, these grid-connected PV systems are distributed throughout the electricity grid. This is the most common type of PV system and hence the focus of this book. There are typically two types of system: commercial and residential.

Commercial systems are generally greater than 10kWp and are located on buildings such as factories, commercial businesses, office blocks and shopping centres. The power generated by these systems is typically consumed by the loads within the building, so no excess power is exported to the electricity grid.

Residential systems refer to those installed on homes and are generally smaller than commercial systems, typically between 1 and 5kWp. The power generated by these systems is first consumed by any loads operating in the house during the day; excess power is fed into the grid providing electricity to nearby buildings.

Figure 1.10 A 40kWp commercial grid-connected PV system located in San Diego, California. The modules have been mounted vertically because the roof space is used as a car park

Source: Global Sustainable Energy Solutions

Figure 1.11 A residential PV system

Source: Green Solar Group

Other solar technology

Solar electric systems are just one of many technologies currently being deployed to harness the sun's power. The other main technologies are described below. It is important to understand that although they all use a form of solar power, each is very different in its operation, design and installation.

Solar thermal power

Rather than using sunlight, solar thermal power uses the sun's heat in a variety of different ways. Solar thermal power is becoming very prominent in both small- and large-scale applications as it is becoming cost-competitive with conventional generation in some applications.

Solar hot water

Solar hot water is one of the most commonly used forms of solar power. Solar power is used to heat water in what is often referred to as a 'solar collector'. Solar collectors are housed on the roof and may also include a rooftop storage tank for the hot water; they should not be confused with solar modules which produce electricity.

There are two main types of solar hot water collectors. The most common collector is the flat plate; a large dark plate collects heat while water flows through tubing beneath. Heat is transferred from the dark plate to the water that, when sufficiently hot, rises to the tank as steam.

Recently, evacuated tubing (a different style of collector) has become popular; this collector uses cylindrical tubing through which water passes. The tubing has an inner and an outer tube and the space between them is a vacuum in order to provide insulation and reduce heat losses from the water to the surroundings. For further reading on solar hot water systems, see Chapter 15.

Figure 1.12 This system includes a solar electric system (back) and flat plate solar hot water system with a storage tank (front)

Source: Green Solar Group

Figure 1.13 A solar hot water system with an evacuated tube collector

Source: Global Sustainable Energy Solutions

Concentrated solar power (CSP)

CSP is a large-scale form of solar thermal power. However, rather than producing hot water, it produces electricity and is effectively a power plant that feeds into the grid. CSP uses large mirrors to concentrate the sun's rays towards a central point (normally a tube). This concentrated energy is used to boil water

Figure 1.14 A CSP installation in the Nevada Desert. Parabolic mirrors concentrate sunlight to heat the water running through the thin tubes in front of the mirrors

and create steam which can be used to drive a steam turbine and generate electricity. CSP systems are most economical when deployed on a large scale and normally act as a power plant; feeding the electricity produced back into the grid. This technology has been in use for many years in the Mojave Desert, California, and is now emerging in other regions such as Spain.

There is a great deal of international interest in CSP technology and a new trend for substituting water with molten salts as a storage medium is emerging. One of the main problems with solar power is that electricity storage is very expensive; however, molten salts are able to store the energy produced by a CSP system as heat for several hours so electricity can still be produced at night time when the sun no longer supplies energy. This form of storage is significantly less expensive than the batteries normally used to store energy produced by photovoltaics.

Passive solar design

Passive solar design refers to designing a building in order to utilize the sun's energy as much as possible. It varies significantly throughout the world, the basic idea being to keep the sun's rays out of the house in the summer and to trap the sun's rays inside the house during winter and so to minimize the need for artificial cooling and/or heating.

- Deciduous trees and vines are useful because if planted in a location exposed to direct sunlight, they will shade the area in summer, but in winter they lose their leaves allowing the sun's rays to come in and warm the area.
- Eaves and pergolas can be designed to shade windows during the summer (when the sun is high in the sky) but allow sunshine in during winter (when the sun is low in the sky) to warm the building. The eaves should also be lightly coloured so they will reflect the sun's rays in summer. If they are a dark colour the material will absorb the sun's rays and heat up.

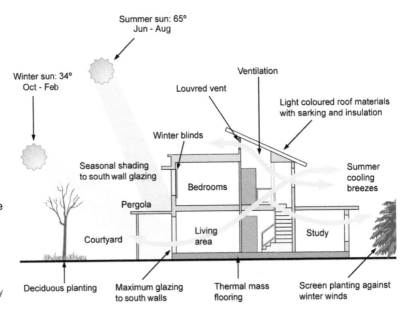

Figure 1.15 A variety of techniques used to capture the power of passive solar: some are very simple while others need to be incorporated into the building's design from the very beginning

Source: Global Sustainable Energy Solutions

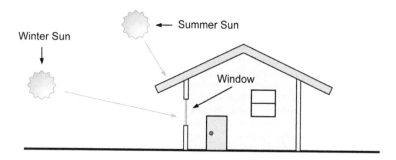

Figure 1.16 When correctly designed eaves can prevent summer sun entering the window but allow winter sun to warm the building

Source: Global Sustainable Energy Solutions

- Windows allow for cooling breezes in summer and can be closed in the winter to seal in warm air. These should be positioned according to local wind direction, i.e. if summer winds commonly blow from the northeast, windows should be placed on the north side of the building and on the east side of the building to capture the breeze.
- Thermal mass and insulation is important for maintaining the building at a comfortable temperature despite fluctuations in outside temperature. This can be as simple as closing curtains in winter to stop heat escaping through the windows.

All these techniques can be used to reduce the need for artificial heating or cooling and reduce the property's energy consumption. They are an excellent complement to a solar electric system as energy-efficient households will use less electricity and so export more solar electricity to the grid.

How to use this book

Getting involved with the solar industry can be daunting. This book provides a clear introduction to the terminology and technology used in grid-connected PV systems. For homeowners and property developers looking to purchase a system, this book gives a clear outline of what to expect during the design and installation processes. For builders and electricians looking to work in the solar industry, it serves as an introduction. It should not, however, be used as a substitute for an accredited training course. Grid-connected PV systems should only be designed and installed by professionals accredited in the country of installation; in all cases local codes and national standards must be consulted. The following chapters outline topics relevant to grid-connected solar electric systems.

Solar Resource and Radiation (Chapter 2)

Before the sun's power can be harnessed by PV devices it is important to establish how much solar energy is available at a given location. This chapter outlines the basics of solar resource data: where it can be acquired and how it is used. Solar resource varies significantly depending on the time of day, location and time of year. All these factors must be taken into account when designing a PV system.

PV Industry and Technology (Chapter 3)

What are PV devices and how are they manufactured? This chapter explores the basic science of PV technology and the many different varieties of PV technology available, including their strengths and weaknesses. It also includes important information about buying modules, such as which standards and certifications to look for and what to expect from a PV module warranty.

PV Cells, Modules and Arrays (Chapter 4)

In order to design and install a PV array a sound knowledge of the electrical characteristics of PV cells is required. This chapter explores how PV cells are connected to make modules and the different ways in which these modules can be connected to create arrays. Sample calculations demonstrate how the voltage and current outputs of a solar array are determined.

Inverters and Other System Components (Chapter 5)

As well as the PV array there are many other components required to make a PV system. The inverter converts the DC power produced by the PV array to AC power appropriate to be fed into the grid. Many different types are available and their different uses are discussed. This chapter also discusses the cabling used to connect the components, PV combiner boxes and module junction boxes which are used to bring together wires and cables. Protection and disconnect devices that ensure the safety of the system, lightning protection, the meter through which electricity flows into the grid and monitoring equipment that provides system owners with useful information regarding power output are also covered.

Mounting Systems (Chapter 6)

Mounting systems are used to secure the PV array to a surface; both roof and ground mounting systems are discussed in Chapter 6. There are many different ways to mount a PV array and the final decision on which system to use is often based on a compromise between performance and aesthetics. This chapter describes the different systems and explores these trade-offs.

Site Assessment (Chapter 7)

Once a property owner decides to purchase a PV system the next step should be a thorough site assessment. The designer will assess important physical features of the site such as size and solar resource in order to determine the optimal system design.

Designing Grid-connected PV Systems (Chapter 8)

Following a successful site assessment the system must be designed. This chapter explores important design processes such as responding to the design brief and choosing system components by referring to local codes, calculation, software and analysis of local conditions.

Sizing a PV System (Chapter 9)

This chapter demonstrates the calculations required to match the inverter and array and ensure they operate safely together. It also shows how to calculate the PV system's yield, accounting for typical system losses. Two sample calculations show the complete process.

Installing Grid-connected PV Systems (Chapter 10)

The installation of PV systems is described with reference to each individual component and the chapter concludes with a discussion of the requirements related to connecting a PV system to the grid.

System Commissioning (Chapter 11)

This chapter describes the process of turning the PV system on and feeding electricity into the grid. Safety is paramount to this process, which involves a thorough visual inspection of the PV system followed by electrical tests and finally the commissioning of the system (that is its connection to the grid). The process is heavily dictated by local codes. However, a rough description of the inspections and tests that may be required is given.

System Operation and Maintenance (Chapter 12)

After commissioning, many PV systems operate without trouble for decades. However, regular maintenance checks are important to ensure the system always operates at its best. This chapter outlines common maintenance practices and who they should be completed by. If a problem does occur, troubleshooting will be required; this can be difficult and should always be done by an accredited professional.

Marketing and Economics of Grid-connected PV Systems (Chapter 13)

In addition to the technical details of PV systems, familiarity with the economics of PV systems is an important part of any solar professional's job. This chapter outlines the types of financial incentives available throughout the world to PV system owners and where further information on them may be found. It also demonstrates how to calculate the payback time of a PV system using this information.

Case Studies (Chapter 14)

Chapter 14 offers case studies of a variety of grid-connected PV systems throughout the world, demonstrating how the principles outlined here are implemented in real life.

2
Solar Resource and Radiation

Throughout history humankind has held a strong relationship with the sun. Many prehistoric cultures worshipped the sun as a deity and modern scientific thinking confirms that the sun and its energy are essential to life on Earth. Solar power has often been a topic of science fiction and fantasy: Isaac Asimov's short story 'The Last Question' (1956) envisaged all the Earth's power being supplied by a small one-mile wide solar collector orbiting the Earth, launched in 2061, while other stories have explored the use of solar power for long-range space travel. Advances in modern solar technology are bringing such fantasies closer to reality and future technology may only be limited by imagination.

Solar resources

The sun is the source of almost all energy on our planet, either directly as sunlight or indirectly as wind and waves. Even the coal reserves we mine today were once living plants that gained their energy from photosynthesis: the process by which sunlight, carbon dioxide and water are converted to carbohydrates. The sun's seemingly endless energy supply is driven by a process known as nuclear fusion, where atoms of hydrogen combine to form an atom of helium releasing a large amount of energy in the process. The helium atom may then combine with other helium or hydrogen atoms to release even more energy.

The energy produced in the heart of the sun is emitted as electromagnetic radiation. Electromagnetic radiation is emitted in many useful forms including microwaves (as used in microwave cooking), radio waves (used in telecommunications) and visible light. Solar cell designers focus on capturing the energy carried in visible light.

Box 2.1 Power and energy

The unit of electric power is the watt (W). Power in watts is equal to the product of volts (V) x amperes (amps).

Power (watts) = voltage (volts) + current (amps) $P = V \times I$

Kilowatts (kW) is the term used for large amounts of power; i.e. 1000 watts = 1kW
Energy is defined as the capacity to do work, or power, over time.

Energy (watt-hours) = power (watts) × time (hours) $E = P \times t$

Large amounts of electrical work or energy are expressed in kilowatt-hours[1], that is simply the amount of power multiplied by the time it is used. The solar radiation data unit commonly used is kWh/m^2. This is the amount of solar energy received by a square metre exposed to full sun (1000W of solar radiation) for a period of one hour.

The difference between power and energy is an important concept as energy usage is the basis for determining the output of renewable energy systems for a given period.

Example

A 60 watt light globe is left on for 12 hours. This light bulb will consume 720Wh or 0.720kWh of energy:

60 watts × 12 hours = 720Wh

Figure 2.1 The IKAROS (Interplanetary Kite-craft Accelerated by Radiation Of the Sun) mission was launched on 21 May 2010. This solar powered craft features a large membrane (or sail) of solar cells that produces electricity and receives additional propulsion from solar radiation

Source: Japan Aerospace Exploration Agency (JAXA)

Figure 2.2 The sun releases vast amounts of electromagnetic radiation – photovoltaic devices provide an efficient means of converting this power into electricity

Source: Global Sustainable Energy Solutions

Figure 2.3 White visible light can be divided into many colours; each ray of colour has a different wavelength/frequency and energy content. Violet light has the most energy while red has the least

Quantifying solar radiation

Although radiation emitted from the sun is fairly consistent there is significant variation in the radiation received at the Earth's surface. This is caused by the Earth's orbit (responsible for the seasons), rotation on its own axis (responsible for night and day) and the albedo (covered later) of certain areas. Solar array

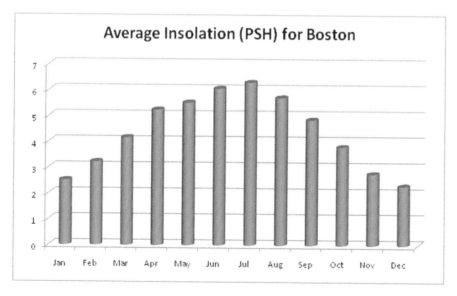

Figure 2.4 Average annual insolation for Boston, MA on a horizontal plane measured in peak sun hours (PSH). Insolation varies widely throughout the year and doubles in summer months so solar module output power greatly increases in summer. This data is for the insolation on a horizontal plane, which is not ideal in a city so far from the equator. Inclining a plane to a tilt equal to the angle of latitude ensures that the plane faces the sun more directly and receives more insolation – optimum angles for PV modules are dealt with in Chapters 7 and 9

Source: Atmospheric Science Data Center, NASA

designers need to be able to quantify how much solar radiation a given site will receive throughout the year. The amount of solar energy received by an area over a day is referred to as insolation and can be measured in kWh/m²/day or peak sun hours (PSH), which are explained below. It is important to note that most PV arrays are tilted at a certain angle above the horizontal and the insolation on the horizontal plane will differ from the insolation on the tilted plane of the array; data is often available for both the horizontal and various tilt angles.

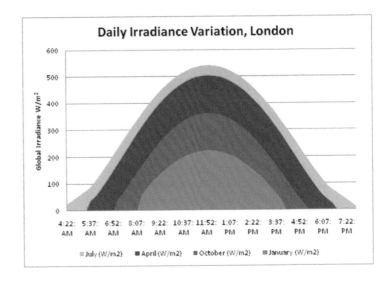

Figure 2.5 Daily irradiance for London, UK, on a horizontal plane measured in W/m² and varying with time of day and month. Solar energy varies throughout the day and peaks in the middle of the day

Source: PVGIS © European Communities, 2001–2008

Box 2.2 Solar radiation terminology

Peak sun hours (PSH): Daily irradiation is commonly referred to as daily PSH (or full sun hours). The number of PSH for the day is the number of hours for which power at the rate of 1kW/m² would give an equivalent amount of energy to the total energy for that day. The terms peak sunlight hours and peak sunshine hours may also be used.

Irradiation: The total quantity of radiant solar energy per unit area received over a given period, e.g. daily, monthly or annually.

Insolation: Another term for irradiation. The amount of solar radiation incident on a surface over a period of time. Peak sun hours (kWh/m²/day) are a measurement of daily insolation.

Irradiance: The solar radiation incident on a surface at any particular point in time measured in W/m².

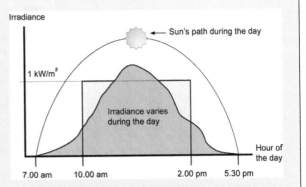

Figure 2.6 Peak sun hours are very useful in system yield calculations (see Chapter 9); one PSH represents 1 hour of radiation at 1kW/m². Because the sun does not shine consistently all day the number of peak sun hours will always be less than the number of hours in a day

Source: Global Sustainable Energy Solutions

Example

If sunlight is received at an irradiance of 1000W/m² for 2 hours, 600W/m² for 1.5 hours and 200W/m² for 1 hour, the total radiation received that day is 3.1PSH:

1000W/m² × 2 hours + 600W/m² × 1.5 hours + 200W/m² × 1 hour = 3100W/m²/day

3100W/m²/day ÷ 1000W/m²/day = 3.1PSH

The effect of the Earth's atmosphere on solar radiation

The Earth's atmosphere reflects a large amount of the radiation received from the sun – without this protection life could not be sustained on the planet. When solar radiation arrives at the top of the Earth's atmosphere it has a peak irradiance value of 1367W/m² (this is known as the solar constant). By the time solar radiation reaches the Earth's surface it has a peak irradiance value of approximately 1000W/m². The difference between the solar constant and the peak irradiance value at the Earth's surface is due to the Earth's albedo – the amount of solar energy reflected from a surface on the Earth at that specific location. Light is reflected from Earth in a variety of ways:

*   Radiation is reflected off the atmosphere back into space.
*   Radiation is reflected off clouds in the stratosphere.
*   The Earth's surface itself reflects sunlight.

The average portion of sunlight reflected from the Earth (the Earth's albedo) is 30 per cent. Polar regions have very high albedo as the ice and snow reflect most sunlight, while ocean areas have a low albedo because dark seawater absorbs a lot of sunlight.

Box 2.3 Key terminology

Direct radiation: Solar radiation that passes directly to the Earth's surface.

Diffuse radiation: Solar radiation that is scattered or absorbed by clouds and gases within the atmosphere and then re-emitted – diffuse radiation is less powerful than direct radiation.

Air mass: The distance that radiation must travel through the atmosphere to reach a point on the surface. This value varies throughout the day for a location.

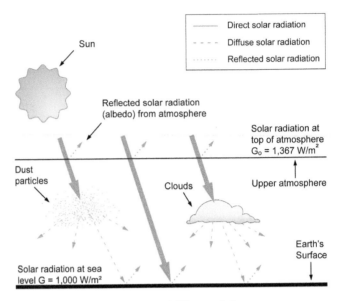

Figure 2.7 Solar radiation comprises direct and diffuse radiation

Source: Global Sustainable Energy Solutions

Irradiance is a combination of direct and diffuse radiation and will depend on the albedo (reflected solar radiation) of that particular location.

That proportion of solar radiation which is scattered, absorbed or re-emitted in the atmosphere is diffuse radiation. Understandably on a sunny day, this scattered diffuse radiation will contribute only to 10 per cent of visible light, but on a cloudy day there will be much more scattering of the solar radiation reaching the Earth's surface which means the amount of diffuse radiation will be much greater. Air mass will also affect the irradiance at a location. The

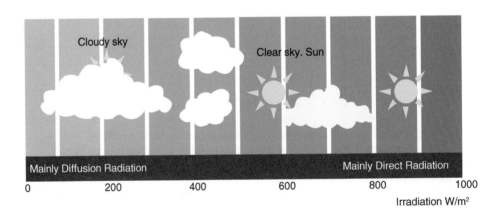

Figure 2.8 The more clouds in the sky the less irradiation there will be and the larger the diffuse radiation component

Source: Deutsche Gesellschaft für Sonnenenergie e.V. (DGS)

greater the air mass, the higher the chance of light being reflected or scattered, meaning there will be less solar radiation reaching the Earth's surface.

Air mass of 1.5 is the standard condition at which solar modules are rated. Air mass zero refers to air mass in space; air mass one corresponds to conditions when the sun is directly overhead. Regions outside the tropics will never experience air mass one, as the sun is never directly overhead.

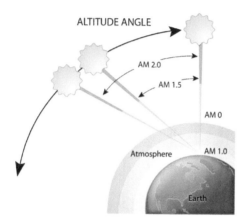

Figure 2.9 From the surface of the Earth, air mass is directly related to the altitude of the sun

Source: Global Sustainable Energy Solutions

Figure 2.10 At sunset the sun is low in the sky and therefore air mass is very high. Light is scattered so much that the white light is separated into its different colours. Most of the colours scatter so that they are not visible but red light scatters the least, which is why we see red sunsets on cloudy days. Blue light is scattered the most, which is why the sky is blue when sunlight passes through the least amount of atmosphere (during the day)

Source: Global Sustainable Energy Solutions

Insolation varies widely depending on location (and time of year) and when designing a system it is very important to consider the solar radiation characteristics for the specific location where the solar array will be installed. For instance, a solar array powering a household in a German city would have to be significantly larger than a solar array powering a household with the same energy consumption in the Australian desert.

Locations far from the equator such as Poland receive a large amount of irradiation during long summer days but very little during winter when the days are very short.

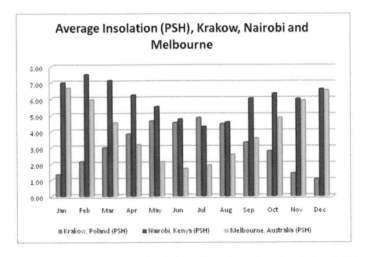

Figure 2.11 Average annual insolation for Krakow, Poland (northern hemisphere), Nairobi, Kenya (near the equator) and Melbourne, Australia (southern hemisphere) on a horizontal plane. This figure shows that areas around the equator (Kenya) receive significantly more solar radiation than areas closer to the poles (Poland and Australia). Areas far from the equator also experience a larger variation in the amount of radiation they receive

Source: PVGIS © European Communities, 2001–2008

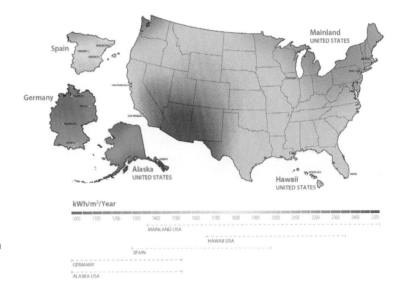

Figure 2.12 This image shows average solar insolation across the US, Germany and Spain. Power output will be significantly higher in areas with higher insolation

Source: Global Sustainable Energy Solutions

Solar radiation data is often available from the national meteorological bureau or may be supplied by the solar module supplier. NASA provides web data for most of the world and the European Commission Joint Research Centre provides a free web tool, Photovoltaic Geographical Information System (PVGIS), that estimates the daily output of a solar array in any location in Europe or Africa (see Chapter 15 for details of these websites).

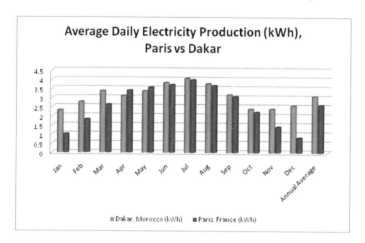

Figure 2.13 Graphed data from PVGIS showing the power output (in PSH) for two identical 1kWp systems, one installed in Dakar, Morocco and the other in Paris, France. Each installation is installed at the optimum tilt angle (34° from the horizontal for Paris and 1° from the horizontal for Dakar). The installation in Dakar produces more electricity because the solar resource is better

Source: PVGIS © European Communities, 2001–2008

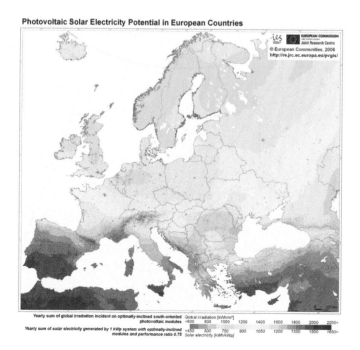

Figure 2.14 This image shows the variation in insolation across Europe: the southern parts of Europe receive significantly more sunlight. Solar systems in these areas will produce more power

Source: PVGIS © European Communities, 2001–2008

Sun geometry

Because of the Earth's orbit and rotation, the position of the sun relative to a solar array is constantly changing. Designers use several geometrical techniques to design an array that will capture the most solar energy possible. The location of the sun is specified by two angles which vary both daily and annually.

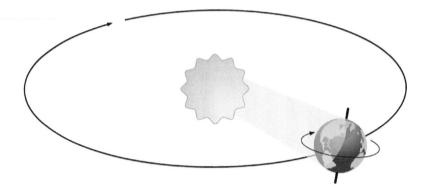

Figure 2.15 Solar resource varies during the day due to Earth's rotation on its own axis and during the year due to Earth's orbit around the sun

Source: Global Sustainable Energy Solutions

Box 2.4 Sun angles

Solar altitude: The angle between the sun and the horizon; the altitude is always between 0° and 90°.

Azimuth: The angle between north and the point on the compass where the sun is positioned. The azimuth angle varies as the sun moves from east to west across the sky through the day. In general, the azimuth is measured clockwise going from 0° (true north) to 359°.

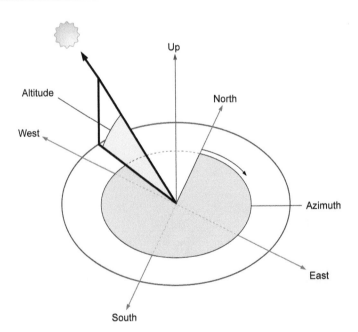

Figure 2.16 The sun's altitude is shown in blue while its azimuth is represented by green

Source: Global Sustainable Energy Solutions

In the northern hemisphere solar arrays are normally installed to face south as the sun is always in the southern sky and in the southern hemisphere solar arrays normally face north. In regions between the Tropic of Cancer and the Tropic of Capricorn this is not always the case, at certain times of year the sun will be in the southern sky for those in the southern hemisphere and in the northern sky for those in the northern hemisphere. In summer the sun will always be higher in the sky than in winter due to the natural tilt of the Earth.

The longest and shortest days of the year are known respectively as the summer and winter solstices and usually fall around 21 June and 21 December. The sun's altitude is highest on the summer solstice and lowest on the winter solstice. The midpoints between the two solstices are known as the equinoxes, normally falling around 20 March and 23 September. Many prehistoric cultures learned to identify and predict the solstices and equinoxes, these dates being useful in planning harvests and religious celebrations.

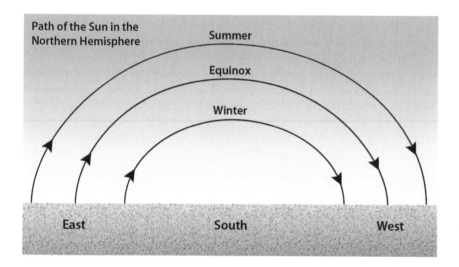

Figure 2.17 Not only does the sun's position vary during the day, it also varies throughout the year and this is an important consideration when choosing the orientation and tilt of an array

Source: Global Sustainable Energy Solutions

Figure 2.18 At sunrise on the summer solstice the sun can be seen to align with the stones at Stonehenge in the UK

Source: Andrew Dunn

The sun's path in the sky for any particular location can be depicted on a two-dimensional surface in a sunpath diagram. This diagram can be used to determine the position of the sun in the sky at any time of the day, for any day of the year. From this information the times at which an area is shaded can be determined, which in turn allows the yearly insolation to be calculated. Practical use of the sunpath diagram is covered in Chapter 7.

The sunpath diagram is composed of:

- azimuth angles, represented on the circumference of the diagram;
- altitude angles, represented by concentric circles;
- sunpath lines from east to west for different dates in the year;
- time of day lines crossing the sunpath lines;
- location information that refers to latitude.

Sunpath diagrams may look entirely different for other areas. Directly on the equator, the sunpath would range equally north and south. Outside the tropics, sunpath diagrams generally look like the one below, though in the northern hemisphere they will be inverted. Note the seasonal variation in the path of the sun. Outside the tropics, the sun only rises to a low altitude in winter and is always north or south (depending on the hemisphere) of the observer. During summer, the sun rises to a higher altitude. The sun will always be farthest north or south at the solstices in all locations, north in June, south in December.

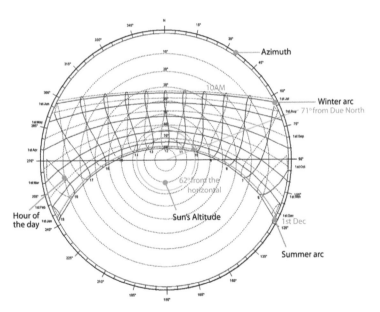

Figure 2.19 Sunpath diagram of Sydney has been used to find the precise location of the sun on 1 December at 10am. 1 December is highlighted on the right of the circle, this line is then followed across until it corresponds with the line for 10am (which follows a figure of eight shape), at the point of intersection the altitude can be found. To determine the azimuth a line must be drawn from the centre of the circle, passing through the intersection between date and time and out to the perimeter; here the azimuth is 71°. Block lines correspond to dates on the left-hand side and dotted lines correspond to dates on the right.

Source: Global Sustainable Energy Solutions

Geometry for installing solar arrays

The position of a solar module is referred to as its orientation. This orientation of the solar array is very important as it affects the amount of sunlight hitting the array and hence the amount of power produced. The orientation generally includes the direction the solar module is facing (i.e. due south) and the tilt angle, which is the angle between the base of the solar module and the horizontal. The amount of sunlight hitting the array also varies with the time of day because of the sun's movement across the sky.

It then becomes apparent that if the sun is overhead and the solar panel is laid flat on the ground, it will capture all the sun's rays, as shown in the image below.

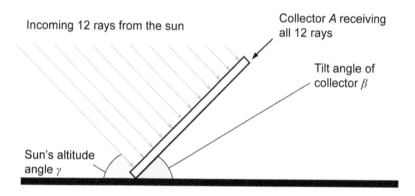

Figure 2.20 Imagine a 1m² solar panel, tilted as shown. At a specific time there are 12 of the sun's rays coming from the sun and hitting the solar panel

Source: Global Sustainable Energy Solutions

Figure 2.21 If the same collector is laid horizontally on the Earth's surface at the same time as in the image above, the collector only captures 9 rays

Source: Global Sustainable Energy Solutions

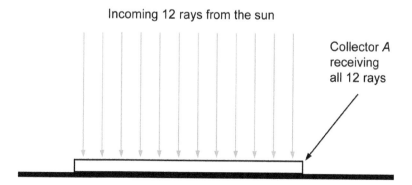

Figure 2.22 A solar array horizontal to the ground will receive the most radiation at solar noon when the sun is directly overhead

Source: Global Sustainable Energy Solutions

Solar modules should be installed so that as much radiation as possible is collected. To achieve this, the solar modules should be installed facing either true south (northern hemisphere location) or true north (southern hemisphere location). There will be some exceptions for installation depending on the local environment (i.e. array's installed in a valley in the southern hemisphere may not necessarily face north). To point a module directly towards the sun at all times would require a solar tracking frame to be installed. This can be expensive, so it is not common practice for most PV applications. However, it will be discussed in further detail in Chapter 6.

Box 2.5 Magnetic declination

When installing a photovoltaic system it is important to consider the magnetic declination of a location (also known as magnetic variation). Magnetic declination is the difference between true north (the direction of the north pole) and magnetic north (the direction in which a compass will point). A solar system should face true north or true south and so the magnetic declination angle of the location should be considered.

In New Orleans magnetic declination is approximately 0° and so the compass would point to true north. In Seattle, however, the magnetic declination angle is approximately 17° east, so when the compass points to 17° east it is in fact facing true north or true north is 17° west. A PV system in Seattle should be installed facing 197° east (or 163° west) so as to face true south.

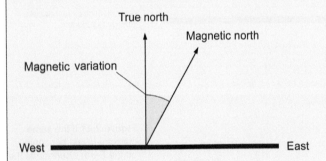

Figure 2.23 Magnetic declination is the bearing between true north and magnetic north

Source: Global Sustainable Energy Solutions

Figure 2.24 In Seattle magnetic north is always 17° east of true north

Source: Global Sustainable Energy Solutions

Note

1 The System International (SI) unit for energy is the joule (J). A joule is a relatively small quantity so large energy quantities such as solar radiation are often expressed in megajoules (MJ). 1MJ is equivalent to 1,000,000 joules. The conversion factor for MJ and kWh is:

$$1kWh = 3.6MJ \text{ or } 1MJ = 1/3.6kWh.$$

3
PV Industry and Technology

Solar cells, also known as photovoltaic (PV) cells, are devices that produce electricity when exposed to sunlight. In 1921 Albert Einstein won the Nobel Prize in Physics for his discovery of the photoelectric effect that is the scientific basis for solar cells. Solar cells were first developed in the 1950s for use in space exploration. They are ideal for this application because they are reliable, need little maintenance (having no moving parts) and require only sunlight, which for this use is in almost limitless supply. In the 21st century solar cells are an increasingly attractive energy source considering the problems posed by greenhouse gas emissions and dwindling fossil fuel energy reserves. They are also popular because they are very versatile and can be used on a small scale, i.e. on homes or to power a single device such as a remote telecommunications repeater station, as well as for large-scale power use.

The potential for photovoltaics has attracted significant investment from governments and businesses all over the world. Globally laboratories are working to improve solar cell efficiencies and to create more powerful cells by further developing existing technologies and by using new materials and manufacturing methods. This chapter explores some solar technologies available today and explains the operation of a PV array.

Semiconductor devices

Solar cells are made from semiconducting materials. Semiconductors are materials that conduct electricity under certain conditions, so they are neither insulators nor conductors. The most common semiconductor material is silicon. To improve its conductivity, it is often combined with other elements in a process known as doping. Semiconductors are used frequently in electronics, including PV cells, light-emitting diodes (LEDs) and microchips such as those used in computers.

Mainstream technologies

Silicon is the most widely used material in solar cell production; most commercial solar cells are made of silicon. Silicon is extracted from silicon dioxide (also known as quartzite); quartzite crystals are quarried and refined to extract silicon for solar cell manufacture. Quartzite crystals are also a major component of sand; however, regular sand is too impure to be used in this process.

Silicon is a versatile material: a metalloid, it exhibits many of the properties of metals (i.e. it is lustrous and solid at room temperature) but also exhibits

Figure 3.1 Semiconductors are used in many electrical components

---

Box 3.1 Solar cell efficiency

Efficiency is a unitless measurement used to indicate how well a device changes one form of energy (i.e. heat, movement, electricity etc.) into another. Efficiency is mentioned a great deal in connection with solar cells, so it is important to understand what is meant by it and what it actually entails in relation to the operation of the cell. Many different forms of efficiency are used to describe the operation of a solar cell:

- Cell efficiency: The amount of electrical power coming out of the cell per amount of light energy that hits the cell. Usually this is measured at standard test conditions (STC): 25°C ambient temperature and 1000W/m² light intensity. A cell will rarely experience STC in the field so it will rarely perform at its rated efficiency. Efficiency and power output are related as follows:

Efficiency = Power OUT/Power IN

The standard value for power in (irradiance) is 1000W/m². If the cell's efficiency is 22 per cent and it has an area of 0.2m² then:

Power OUT = Efficiency × Power IN

Power OUT = 0.22 × 1000W/m² × 0.2m²

Power OUT = 44W

- Module efficiency: The efficiency of a module is measured the same way as that of a cell, the difference being the inclusion of losses from reflection and shading of the glass as well as a few other minor losses.

Purchasers should always be wary when considering efficiency figures; it is important to find out what the test conditions were and to understand how the performance will change when the array is installed, the environment is never at standard test conditions and variation in temperature, wind and insolation will strongly affect the array's performance. Efficiency ratings given at STC are generally best for comparing cells/modules, rather than calculating system yields (covered in Chapter 8), as cell temperatures in the field change throughout the day and may never achieve 25°C. The efficiencies achieved for a PV cell in a laboratory are generally significantly higher than the efficiency of a readily available PV module, because many techniques used in laboratories are not yet economically viable for use in mass production.

some properties of non-metals. There are several different types of silicon solar cell: monocrystalline, polycrystalline and amorphous silicon will be discussed in this chapter.

Monocrystalline silicon

These solar cells are produced from a single silicon seed crystal placed in a crucible of molten silicon and drawn out slowly while rotating. In this manner it is possible to produce a larger pure crystalline silicon ingot, which is then sliced into thinner wafers. Monocrystalline silicon solar cells are the most efficient and generally the most expensive, although the higher initial cost may be justified by their increased power output as they are highly efficient compared to other silicon technologies. The highest recorded efficiency for

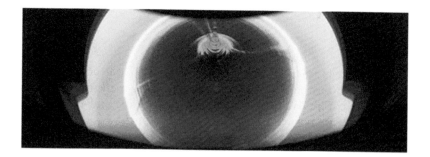

Figure 3.2 A single-crystal silicon ingot is drawn from the molten silicon

Source: Silfex

Figure 3.3 As well as being used in rooftop and grid-connected applications, 22 per cent efficient monocrystalline PV cells were used by Sunswift IV, the winning silicon solar car at the 2009 World Solar Challenge held in Australia

Source: Photo: Daniel Friedman

monocrystalline silicon solar cells is 25 per cent; commercially available solar cells are now being produced with efficiencies of 22.5 per cent and finished modules with efficiencies of 19 per cent.

Multicrystalline/polycrystalline silicon

Multicrystalline or polycrystalline silicon solar cells are manufactured by block casting molten silicon, so they are not made from a single crystal ingot but rather from one composed of many small crystals, which grow in random orientations as the molten material solidifies. This produces lower efficiencies than monocrystalline cells; however, it is still a very popular technique because it is easier and less expensive. Multicrystalline and monocrystalline silicon solar cells are those most commonly used in PV arrays, and commercially available multicrystalline solar cells can now reach laboratory efficiencies over 18 per cent, with the record for module efficiency being 17.84 per cent.

Thin film solar cells

Thin film solar cells are made from materials suitable for deposition over large areas. They need only be about one micron thick, hence the name thin film (a dot such as '.' covers 615 microns and multicrystalline and monocrystalline silicon solar cells are normally about 300 microns thick).

Increasing material prices and high worldwide demand for affordable photovoltaics have led to increasing interest in thin film solar cells. They are less expensive to manufacture than crystalline solar cells and a lot of research has been directed towards increasing efficiencies, the current laboratory record

Figure 3.4
Multicrystalline cells are easily identifiable by their distinct grain structure which gives a glittering effect in sunlight

Source: Global Sustainable Energy Solutions

for thin film solar cell efficiency being 20.1 per cent. However, commercially available thin film modules are between 6 and 12 per cent efficient. Thin film cells are being used more and more frequently in buildings and are often used for gadgets such as solar-powered watches and calculators.

The most common materials are amorphous silicon (a-Si, still silicon but in a different form), cadmium telluride (CdTe) and copper indium (gallium) diselenide (CIS or CIGS). In amorphous silicon the molecules are randomly aligned instead of having a fixed crystalline structure as seen in mono- and poly-silicon. CdTE, CIS and CIGS are all polycrystalline materials, their internal structure being similar to that of polycrystalline silicon; however, they are very different materials. Thin film solar cells are well suited to high volume manufacturing; they are made using the chemical vapour deposition (CVD) process, where the material is deposited onto large area materials, e.g. coated glass, flexible plastic or stainless steel sheet.

Figure 3.5 Amorphous silicon modules are easily distinguished from crystalline modules by their dark and uniform colour but they also look like non-silicone thin film modules

Table 3.1 Comparison of different solar technologies

Cell material	Module efficiency	Surface area required for 1kWp in metres squared	Surface area required for 1kWp in square feet
Monocrystalline silicon	14–20%	5–7m²	54–77 ft²
Polycrystalline silicon	13–15%	6.5–8.5m²	72–83ft²
Amorphous silicon thin film	6–9%	11–16.5m²	110–179ft²
CdTe thin film	9–11%	9–11m²	98–110ft²
CIS/CIGS thin film	10–12%	8.5–10m²	90–108ft²

Source: IEA

The size of the array required is largely dependent on the efficiency of the module, although the most efficient module may not always be the most desirable. Less efficient modules are generally less expensive and may be most cost-effective in areas where the amount of space available for the array is not a constraint.

Contacts

For either PV cells or modules to work, they need electrical connections. A solar cell has electrical connections provided by metal conducting strips that collect electrons produced in the cell and allow this current to flow through the circuits. These metal strips are known as contacts and there are several different contact technologies currently in use.

The most common way to manufacture solar cells is by screen printing, where the metal is simply printed onto the cell. This method is very reliable and is typically applied to solar cells producing efficiencies of about 12–15 per cent. Several factors need to be balanced when using screen printing:

- If there is too much space between the contacts, the cell will be less efficient.
- If the area covered by the contacts is too large, the cell will receive less sunlight and produce less power.
- For the contacts to be effective, the top of the cell often needs to be treated in such a way that its absorbance of high-energy blue light is reduced.

Another method used by manufacturers to achieve solar cell efficiencies over 20 per cent is rear or back contacts. This technology increases the working cell area, allows a simplified automated production and the cell wiring is hidden from view. The most efficient commercially available silicon solar cells use rear contacts (18–23 per cent). There is no metal on the front of the cell, which means the whole cell is producing electricity.

The highest efficiency silicon cells (25 per cent laboratory efficiency) derive from another method known as buried contact solar cells (BCSC). This method uses small laser grooves which are cut in the cell and the metal is inserted into

Figure 3.6 The metal contacts on this polycrystalline cell are clearly visible

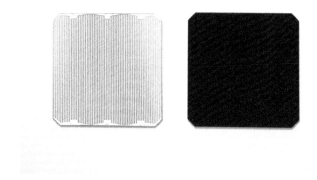

Figure 3.7 Front and back of SunPower rear (back) contact monocrystalline silicon solar cell

Source: SunPower

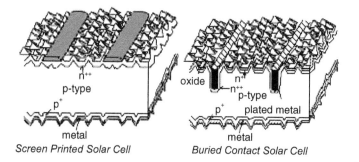

Screen Printed Solar Cell

Buried Contact Solar Cell

Figure 3.8 BCSC technology is used to manufacture solar cells with much higher efficiencies than those that use screen printing. However, screen printing is still the dominant form of contact manufacturing because of its technical simplicity and cost effectiveness

Source: School of Photovoltaic and Renewable Energy Engineering, University of New South Wales

the groove. Buried contact and rear contact technologies are advantageous because they solve many of the problems associated with screen printing. Buried contacts cover very little surface area because most of the contact is inside the groove. This means they don't block as much sunlight and the contacts can be very close together because they are so thin. The sides of the grooves are treated instead of the top, so the PV cell's absorption of blue light is not affected. Rear contacts also eliminate many of these problems, because they are located at the back of the cell, they do not shade the cell and so the number of these cell contacts can be increased to improve cell efficiency.

Buying solar modules

Many different PV modules are available worldwide. It can, however, be difficult to determine whether the products and suppliers are reliable. It is important to be able to ensure that the PV modules and the electrical installation are safe and of good quality. Poor practice and sub-quality products pose serious risks to the safety and financial viability of this investment. Outlined below are factors to consider when purchasing a PV system, bearing in mind that a good system should not only physically last for the duration of the warranty but also work effectively during that time. This means that the modules, additional equipment (discussed in Chapter 5), mounting frame (discussed in Chapter 6) and roof must all last as long as the modules or be replaced during that time.

Solar modules are sold with a lengthy manufacturer's warranty, e.g. 25 years for 80 per cent of output. While the guaranteed output must be included in any system output estimates, the buyer needs to understand that if they are installing a product with such a potentially long warranty, the structure (e.g. the building and roof itself) and other system components need to last a similar length of time. For some equipment, i.e. the inverter, a replacement may be planned over the life of the PV system.

Standards

The PV industry is growing rapidly, resulting in many new manufacturers producing PV modules. It is important that only quality modules are installed and standards do exist. The most common standards applicable to PV modules are:

- IEC 61215 Crystalline silicon terrestrial photovoltaic (PV) modules – Design qualification and type approval.
- IEC 61646 Crystalline thin-film terrestrial photovoltaic (PV) modules – Design qualification and type approval.
- IEC 61730 Photovoltaic (PV) module safety qualification – Requirements for construction and requirements for testing.

These standards originate from the International Electrotechnical Commission: www.iec.ch. In many countries a PV module must evidence compliance with either IEC 61215 or IEC 61646 (depending on whether it is thin film or crystalline silicon technology) and IEC 61730.

Certifications

Quality modules should comply with the IEC standards but may also adhere to local standards. Certifications generally ensure that modules adhere to all relevant standards in addition to the specific photovoltaics standards; examples range from wind loading to salt mist corrosion resistance (IEC 61701). The two most common certifications are:

- CE marking: The CE mark stands for conformité européenne. By using this mark the manufacturer is declaring that their module complies with the essential requirements of the relevant European Union health, safety and environmental protection legislation. The CE marking is mandatory on modules sold and installed in the EU.
- Underwriters Laboratory (UL): The Underwriters Laboratory is a US-based testing facility. Modules bearing a UL mark have been tested for compliance to IEC standards and local standards (depending on the country code next to the UL mark). Further information can be found in Chapter 15.

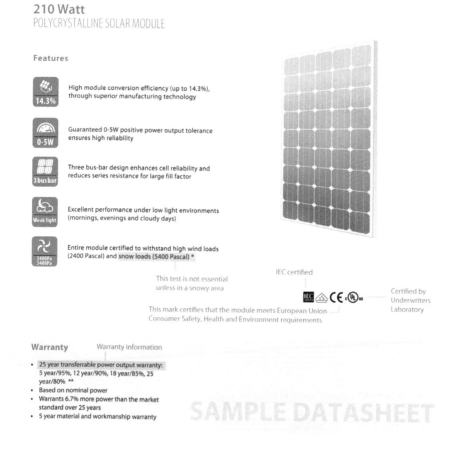

210 Watt
POLYCRYSTALLINE SOLAR MODULE

Features

High module conversion efficiency (up to 14.3%), through superior manufacturing technology

Guaranteed 0-5W positive power output tolerance ensures high reliability

Three bus-bar design enhances cell reliability and reduces series resistance for large fill factor

Excellent performance under low light environments (mornings, evenings and cloudy days)

Entire module certified to withstand high wind loads (2400 Pascal) and snow loads (5400 Pascal) *

This test is not essential unless in a snowy area

IEC certified

This mark certifies that the module meets European Union Consumer Safety, Health and Environment requirements

Certified by Underwriters Laboratory

Warranty   Warranty information

- 25 year transferrable power output warranty: 5 year/95%, 12 year/90%, 18 year/85%, 25 year/80% **
- Based on nominal power
- Warrants 6.7% more power than the market standard over 25 years
- 5 year material and workmanship warranty

SAMPLE DATASHEET

Figure 3.9 Sample data sheets showing information regarding the technical and safety standards to which the sample module complies. All manufacturers' data sheets are specific to their products only, and when accessing any solar module specifications (either in soft or hard copy) the manufacturer should be contacted to confirm that their published material is current.

Source: Global Sustainable Energy Solutions

Warranties

PV modules are typically supplied with three levels of product warranty:

1 1-, 2- or 3-year warranty on the physical manufacture of the module itself, i.e. the frame, encapsulant, glass, module junction box etc.;
2 10–12 year warranty that the module will produce 90 per cent of its rated output;
3 20–25 year warranty that the module will produce 80 per cent of its rated output.

If any of these conditions are not met, the question of a warranty claim will depend on the manufacturer's published warranty and/or the commercial laws of the country where the module was purchased. For example, if someone buys a module from overseas and imports it for sale or their own use, then it is assumed that they understand that they have to send the module back to where they purchased it to claim warranty.

If an imported module is purchased from a 'reseller' of the product, i.e. not the manufacturer itself or their agent, the reseller might have an arrangement with the manufacturer that all warranties are handled by the manufacturer directly or, in some countries, the importer is deemed to be the manufacturer, so all warranties rest with them.

PV modules are extremely reliable and stable products, but it is still advisable to find out the warranty provisions of any product being considered for purchase and installation.

Emerging technologies

Besides well-established mainstream technologies many new technologies are emerging. Some build on existing knowledge and research while others incorporate innovative new materials and techniques.

Dye-sensitized solar cells

Dye solar cells are still technologically immature. At the atomic level they operate very differently from other solar cells and do not use silicon. Dye solar cells use titanium dioxide (also used in toothpaste) and coloured dyes; they can be manufactured at a much lower cost than other solar cells and work better in low light. Dye solar cells are transparent and can be produced in many different colours, making them ideal for architectural applications as windows. Dye solar cells also have potential in military applications as they can be made in camouflage patterns.

Figure 3.10 Dye solar cells are available in a range of attractive colours and are transparent so they can be integrated into the facade of a building

Source: Dyesol

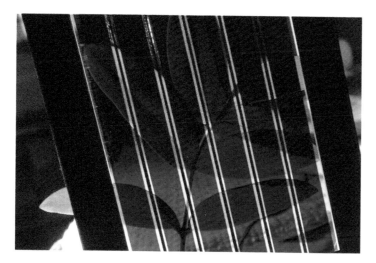

Figure 3.11 Dye solar cell

Source: Dyesol

Currently the highest cell efficiency achieved in a laboratory is 12 per cent and cells with an efficiency of 7 per cent are in production.

Sliver cells

Sliver cells were developed at the Australian National University and are very thin monocrystalline silicon solar cells. They are unique as silicon cells because they are bifacial (they can absorb light from both directions). Sliver technology has achieved cell efficiencies of over 19 per cent and module efficiencies of 13.8 per cent. The technology is in its early stages of commercialization but shows a lot of potential for applications in building-integrated photovoltaics.

Figure 3.12 Sliver cells are also transparent and flexible

Source: Australian National University (ANU)

Heterojunction with intrinsic thin layer (HIT) photovoltaic cells

HIT modules use both crystalline silicon solar cells and amorphous silicon thin film technology and have achieved module efficiencies of 17 per cent and cell efficiencies of 22 per cent.

III-V Semiconductors

III-V or extrinsic semiconductor solar cells use an element from group III of the periodic table and an element from group V (as opposed to silicon which is a group IV element), such as gallium arsenide which is commonly used in space grade solar cells. These solar cells are commonly multi-junction so they are in fact many layers of solar cells, which will collect different colours of visible light. They also frequently use advanced solar concentrator technology to maximize incoming solar radiation. Extrinsic semiconductor multi-junction solar cells are the most efficient and most expensive technology on the market. The highest recorded laboratory efficiency is 41.6 per cent and is held by Spectrolab Inc in the US. Due to their high cost, III-V semiconductor cells are normally used for space applications such as satellites or other big-budget, high-performance, solar-powered devices such as solar planes and solar racing cars (see Figure 3.13).

Solar concentrators

Solar concentrators are used to increase the intensity of light hitting the cell so that it will generate more electricity (the output power produced by a solar cell is dependent on the intensity of light hitting that cell). There are many different kinds of solar concentrators available for a variety of applications, but the most common are lenses or reflective troughs used to focus light.

Solar concentrators are advantageous because they increase the power output so that the system requires fewer solar cells (solar cells are always the

Figure 3.13 Nuon Solar Team from the Netherlands races its solar car Nuna5 in the Suzuka Dream Cup 2010, a track race held in Japan

Source: Hans Peter van Velthoven

largest cost when installing a system). Many of these cells require a cooling system to perform well, as cell temperatures can get very high. Challenges to this technology involve developing concentrators that are sturdy and reliable enough to survive decades in solar installations, sometimes in harsh conditions such as desert, and to keep costs as low as possible.

Figure 3.14 The mirrors either side of the module are used to concentrate light and increase power output

4

PV Cells, Modules and Arrays

In the previous chapter the many different types of PV cells and manufacturing techniques were discussed. This chapter explores how PV cells are used to create PV modules that can then be used to create a PV array, which is the principle component of a grid-connected PV system.

Box 4.1 Electrical basics

A basic knowledge of electrical terminology is essential to understanding the operation of a PV cell.

Current: represented by the symbol I, it is measured in amperes (A or amps). Current is produced by the flow of electrons; the higher the current the higher the rate of flow. There are two types of current:

AC: Alternating current is so named because the direction of the electron flow changes: electrons first flow one way and then the other – this switching continues at a constant frequency. Mains power uses AC current.

DC: Direct current does not switch but rather flows steadily in one direction. PV cells produce DC current.

Voltage: Always measured across two points, it is the change in potential energy per unit charge between those two points. It is represented by the symbol V and it is measured in volts (V).

Energy: Energy is measured in watt-hours (Wh) or joules (J) and is a measure of the ability to do work. A person eating a biscuit gains from that food energy that they can expend in doing work, e.g. walking up a flight of stairs. Kilowatt-hours (kWh) are commonly used to describe electrical energy produced by a PV system.

Power: Power is measured in watts (W) or joules/second (J/s) and is the rate at which energy is supplied. 1 watt is equivalent to 1 joule per second. Power is the product of current (I) and voltage (V):

$$P = I \times V$$

Circuit: A circuit is the system of wires and electrical components (including PV modules) through which current flows. Current can only flow through a closed circuit.

Series connection: Two elements of a circuit are connected in series when they are connected one after the other, such that the current travels through them equally, while the voltage is divided between them (the largest voltage will occur across the largest resistance).

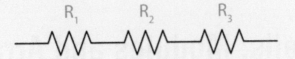

Figure 4.1 Resistors connected in series

Source: Global Sustainable Energy Solutions

Parallel connection: Two elements of a circuit are connected in parallel when they are connected across the same potential difference (i.e. the same voltage), and the current is divided between them.

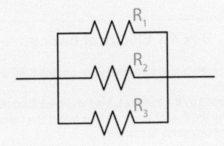

Figure 4.2 Resistors connected in parallel

Source: Global Sustainable Energy Solutions

Characteristics of PV cells

Each type of PV cell is unique and has its own individual characteristics; it is very important to be aware of this when designing an array as the characteristics of the cells in one type of module can affect the power output of the other modules connected to it.

Box 4.2 Key terminology

Open circuit: A circuit is broken and no current can flow.

Short circuit: There is a connection across two terminals of a power source, so current can flow freely between them; this is usually due to a fault in the circuit. Voltage in short-circuit conditions is zero.

Open-circuit voltage V_{oc}: The voltage measured across a PV cell under open-circuit conditions, current is zero.

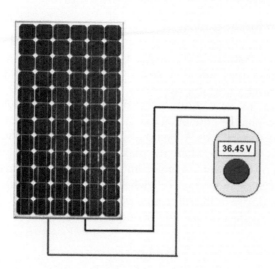

Figure 4.3 Measuring the open-circuit voltage of a module using a multimeter

Source: Frank Jackson

Short-circuit current I_{sc}: The current measured across a PV cell under short-circuit conditions, voltage is zero.

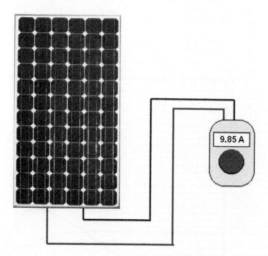

Figure 4.4 Measuring the short-circuit current of a module using a multimeter

Source: Frank Jackson

Maximum power point: The theoretical maximum power output of a PV cell. The maximum power point (P_{max}) is the product of the maximum power point voltage (V_{mp}) and the maximum power point current (I_{mp}).

Graphic representations of PV cell performance

It is very common for the features of a PV cell to be represented graphically as a current-voltage (or I-V) curve. An I-V curve tracks the PV cell's performance and highlights key features such as V_{oc}, I_{sc} and P_{max}. A PV cell will always operate along this curve, i.e. at a given voltage; the current produced will always have the same value and vice versa.

A power curve is used to find the maximum power point. A power curve plots the voltage along the horizontal axis and the power (current multiplied by voltage) along the vertical axis. When this is superimposed on the I-V curve for the same cell, it is very clear where the maximum power point lies.

The I-V and power curves are important because it is necessary to know the characteristics of each individual cell when designing a module. Connecting cells with dramatically different characteristics together will have a large (generally negative) effect on the power output of the PV module.

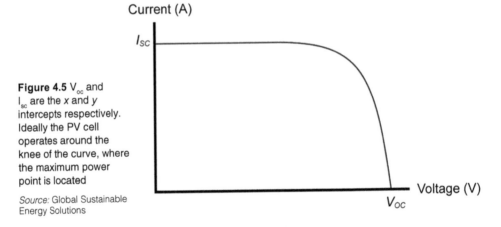

Figure 4.5 V_{oc} and I_{sc} are the x and y intercepts respectively. Ideally the PV cell operates around the knee of the curve, where the maximum power point is located

Source: Global Sustainable Energy Solutions

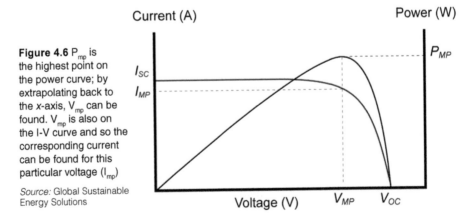

Figure 4.6 P_{mp} is the highest point on the power curve; by extrapolating back to the x-axis, V_{mp} can be found. V_{mp} is also on the I-V curve and so the corresponding current can be found for this particular voltage (I_{mp})

Source: Global Sustainable Energy Solutions

Photovoltaic modules and arrays

Figure 4.7 Single photovoltaic module; the module's 36 cells can be seen arranged in a 4 x 9 grid

Source: Global Sustainable Energy Solutions

Figure 4.8 View beneath a PV array

Source: Global Sustainable Energy Solutions

Figure 4.9 PV array composed of monocrystalline modules

Source: Global Sustainable Energy Solutions

Connecting PV cells to create a module

PV cells of identical characteristics are wired together in series to create a module. In series, their voltages will add while the current remains constant, i.e. the current of the module is equal to the current of one cell. Power losses occur when a cell is underperforming. It may be damaged or more commonly shaded and this will be discussed later in this chapter.

The characteristics of cells are given on the manufacturer's data sheet, freely available on the manufacturer's website; these are used to design a PV array. Designers should always contact the manufacturer to make sure the data sheet on their website is current.

Specification sheets

PV modules purchased from reputable manufacturers should come with specification sheets (also known as data sheets). A data sheet includes important technical information required to design and install a PV array. It may also be useful for a consumer to look at the data sheet when comparing different modules as it provides basic information about efficiency, rated power and physical size.

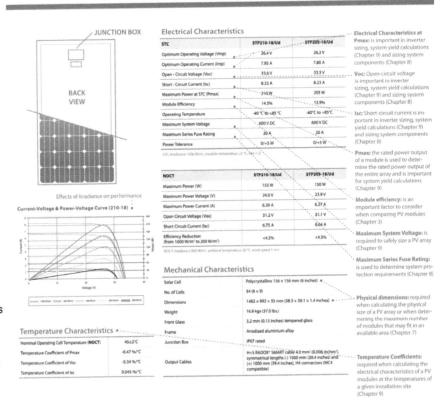

Figure 4.10 Sample data sheet. Data sheets include important technical information such as electrical and mechanical characteristics. All manufacturers' data sheets are specific to their products and when accessing any solar module specifications (in either soft or hard copy) the manufacturer should be contacted to confirm that their published material is current

Source: Global Sustainable Energy Solutions

Box 4.3 Standard test conditions (STC)

These are defined as the conditions under which all modules are tested and specifications given so that comparison between different cells and modules is possible. These are:

- Cell temperature 25°C
- Irradiance 1000W/m²
- Air mass 1.5

Creating a string of modules

A string comprises a number of PV modules connected in series. The electrical characteristics of PV modules connected in series to form a string are the same as PV cells connected in series to form a module: meaning the output voltage of the string will be the sum of the output voltages of all the modules and the output current of the string will be the lowest output current of any module.

Modules can also be connected in parallel. In this case the current output of the modules will add instead of the voltage. The output voltage is that of a single module.

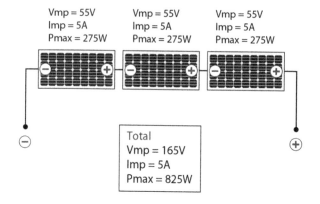

Figure 4.11 When three identical modules are connected in series to form a string their voltages add and the total current is that of one single module. The power output of the string is calculated using P = I x V

Source: Global Sustainable Energy Solutions

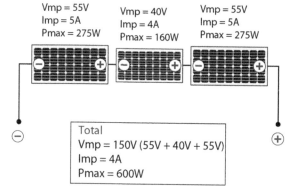

Figure 4.12 When non-identical modules are connected in series the voltages will still add; however, the current of the string will be the lowest current of any single module (in this case 4A). The power output of the string is then calculated using P = I x V

Source: Global Sustainable Energy Solutions

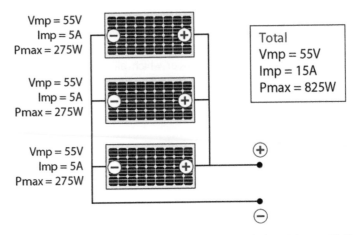

Vmp = 55V
Imp = 5A
Pmax = 275W

Vmp = 55V
Imp = 5A
Pmax = 275W

Vmp = 55V
Imp = 5A
Pmax = 275W

Total
Vmp = 55V
Imp = 15A
Pmax = 825W

Figure 4.13 Three identical modules are connected in parallel; the total current is the sum of each individual current, while the total voltage is the voltage of a single module. The power is once again calculated using P = I x V

Note: The power has the same value as for three identical modules connected in series.

Source: Global Sustainable Energy Solutions

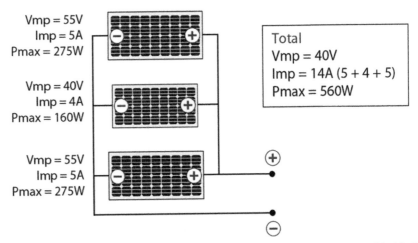

Vmp = 55V
Imp = 5A
Pmax = 275W

Vmp = 40V
Imp = 4A
Pmax = 160W

Vmp = 55V
Imp = 5A
Pmax = 275W

Total
Vmp = 40V
Imp = 14A (5 + 4 + 5)
Pmax = 560W

Figure 4.14 When non-identical modules are connected in parallel the currents add while the output voltage is equal to the lowest single module voltage. The power output of the modules is then calculated using P = I x V

Note: This configuration produces even less power than if the same three modules were connected in series.

Source: Global Sustainable Energy Solutions

Creating an array

Array designers will connect PV modules using a combination of series and parallel to produce the output current and voltage suitable to a market

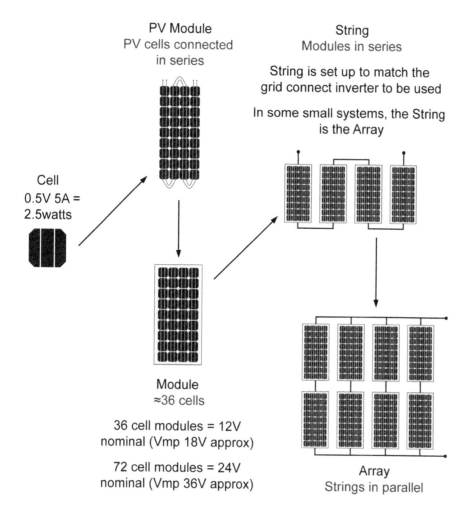

PV Module
PV cells connected
in series

String
Modules in series

String is set up to match the
grid connect inverter to be used

In some small systems, the String
is the Array

Cell
0.5V 5A =
2.5watts

Module
≈36 cells

36 cell modules = 12V
nominal (Vmp 18V approx)

72 cell modules = 24V
nominal (Vmp 36V approx)

Array
Strings in parallel

Figure 4.15 The process of making an array; starting from cells, modules are created, then connected in series to form strings and finally strings are connected in parallel to make an array

Source: Global Sustainable Energy Solutions

application. Modules are typically connected in series to form a string and these strings are connected in parallel to form an array.

As discussed earlier in the chapter, PV arrays produce DC power while the mains require AC power. An inverter is required to convert the DC solar power to AC power and the array is wired in such a way that the maximum power point voltage of the array lies within the range of the grid interactive inverter (inverters are discussed further in Chapter 5).

Photovoltaic array performance

The performance of a PV array is affected by a variety of factors: the most significant of these (temperature, irradiance and shading) are discussed here.

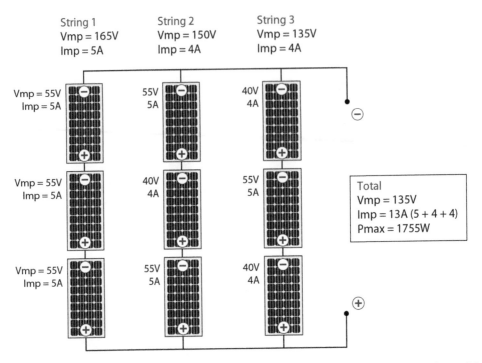

Figure 4.16 To calculate the power output of an array, first calculate the output of each string (as explained previously) and then treat the strings as though they were modules and calculate the power output by adding them in parallel

Source: Global Sustainable Energy Solutions

Irradiance

The amount of solar radiation (sunlight) hitting the cell will largely determine its power output.

The output of a PV array can be estimated using performance data provided by the manufacturer on the data sheet. All arrays have a rated peak power output, i.e. an array can be described as a 1.5kWp array – meaning that PV is installed to provide a 1.5kW peak of power. This output has been determined by the manufacturer using standard test conditions. Using this information and local solar insolation data (see Chapter 2), it is possible to estimate the output of an array.

Example

On a clear sunny day a 2kWp PV array received 6 peak sun hours: the 6 peak sun hours equate to an energy input of 6000W/m² per day. Expected output can be determined as follows:

peak power output × peak sun hours = expected output

2kW × 6PSH = 12kWh

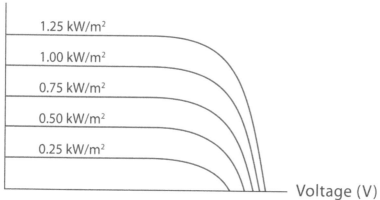

Figure 4.17 The I-V curves for a cell operating at different irradiance values show the increase in power output with irradiance

Source: Global Sustainable Energy Solutions

The PV array produced 12kWh for that day (before any power loss factors are accounted for). There will be varying insolation data available on a monthly basis, so the whole year's output from the PV array can be estimated from the monthly data. However it is important to note that this technique is never used in practice because a PV system will rarely experience STC. Chapter 9 demonstrates how designers account for the power losses that occur when the modules are not operating at STC.

Table 4.1 Monthly peak sun hours for Sydney, Australia

Month	Average peak sun hours (31° tilt) per day
January	5.38
February	5.11
March	4.84
April	4.42
May	3.87
June	3.90
July	4.04
August	4.68
September	5.33
October	5.51
November	5.44
December	5.57
Average	4.84

Source: NASA

Annual average: 4.84 peak sun hours/day

A 2kWp PV array would produce an average of 2kW × 4.84PSH/day = 9.68kWh/day. This results in an average of 9.68Wh/day × 365days/year = 3533.2kWh/year.

Temperature

Not only does the solar radiation hitting the modules produce electricity, it also heats up the modules. It is not uncommon for a PV module to reach 70°C on a sunny day in a temperate climate. As temperature increases, the open-circuit voltage decreases rapidly while the short-circuit current increases slowly. Power output is voltage multiplied by current and so will decrease as well. When designing systems, engineers will often use the following approximation (depending on local design codes and guidelines):

cell temperature = ambient temperature + 25°C

As hot temperatures adversely affect power output, output from a PV array has to be calculated taking the temperature effects into consideration, i.e. derating an array's output based on the operating temperature conditions. Likewise, as

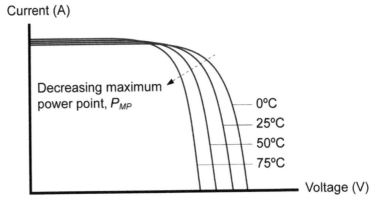

Figure 4.18 As power = current x voltage (I x V), as voltage decreases, power decreases

Source: Global Sustainable Energy Solutions

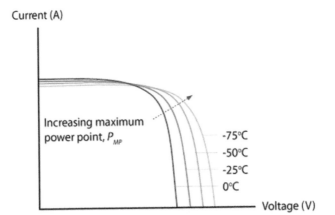

Figure 4.19 In areas that experience extremely cold temperatures, the voltage increases as the ambient temperature decreases

Source: Global Sustainable Energy Solutions

Figure 4.20 PV modules perform well in cold and sunny areas. They are often used as a power source in Antarctica during the summer. However, because there is so little light in winter, diesel generators are typically used as back-up

Source: public domain/Wikipedia/US Antarctic Expedition

cold temperatures can increase the power output due to the voltage increase, the maximum voltage threshold of the system needs to be accurately calculated to ensure that this voltage cannot exceed the inverter's ratings. These techniques are discussed in Chapter 9.

The installation of a PV array can directly affect the operating temperature of the array itself. One contributing factor is when an array is installed flush to a roof surface – meaning that there is limited air flow across the back of the modules to moderate the module temperature. This can have a doubly negative affect because the roof itself will give off heat, particularly a tin roof, and the module will retain

Figure 4.21 Vegetation, chimneys, buildings, dirt and snow can all shade a PV module

Source: Global Sustainable Energy Solutions

heat on the underside, so forced ventilation across the roof surface and behind the modules is often necessary. If the installation cannot include both methods of ventilation, the output of the array will have to be derated to reflect negative aspects of the installation. Ventilation is discussed in more detail in Chapter 6.

Shading

PV cells require sunlight in order to produce electricity. If a cell receives no sunlight due to shading it will not produce any power (even a small area of cell

Figure 4.22 Even this small shadow can reduce the amount of electricity a module produces – a small shaded area can, under certain circumstances reduce module output by 80–90 per cent as well as affecting the rest of the array

Source: Global Sustainable Energy Solutions

Figure 4.23 The discoloured cell in this array was caused by hot spot heating

Source: Global Sustainable Energy Solutions

shading can result in a large reduction in power output). Cells in modules are normally connected in series, so when one or several cells are shaded, the current output of the module will be reduced. If the module is part of an array, then the current output of the array will also be reduced. This will also occur if a cell is damaged and unable to produce power.

Shading of the array can lead to irreversible damage. Hot spot heating occurs when a cell is shaded such that its power output is reduced and most of the current being produced by the other (unshaded) cells is forced through that one cell causing it to heat up. This often leads to cell damage (cracking) and can also damage the glass encapsulation.

It is difficult to prevent shading. However, diodes can be used to mitigate temporary shading (i.e. leaves that may have fallen on the array). When a cell is shaded or damaged, a diode can be used to give current another path to follow. It will skip the damaged or shaded cell completely and have minimum impact on the power output of the array. This kind of diode is referred to as a bypass diode and manufacturers typically install one, two or three bypass diodes per module.

5

Inverters and Other System Components

In addition to the PV modules, a PV system requires other components in order to interact effectively and safely with the power grid. This chapter discusses inverters, PV combiner boxes and metering (net and gross). PV arrays are covered in Chapter 4 and the PV array frame is discussed in Chapter 6.

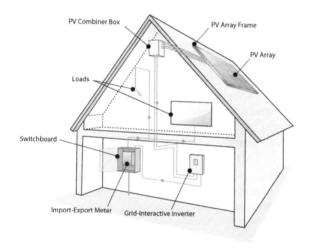

Figure 5.1 The primary system components are shown in this diagram (other configurations are possible). This is a net-metering arrangement where the electricity generated by the PV array and converted to AC by the grid-interactive inverter is either used on site or exported to the grid

Source: Global Sustainable Energy Solutions

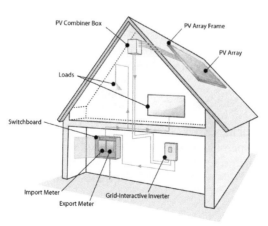

Figure 5.2 Alternative PV system using a gross-metering arrangement where all the power generated by the PV array is exported to the grid and all the power required by the loads is imported from the grid

Source: Global Sustainable Energy Solutions

Inverters

Photovoltaic arrays produce DC, while the typical electricity grid is AC and most electrical devices operate on AC power. To ensure that the power produced by a PV array will flow into the grid, it is necessary for an inverter to convert the DC power produced by the PV array to AC power.

The circuit design of inverters makes this conversion possible: the alternating current is created by the inverter's switching mechanisms, which rapidly open and close the circuit. Transformers, discussed later in the chapter, are then used to increase the voltage to the level required by the grid.

There are two main types of inverters: battery or power inverters, which use batteries as their power source, and grid-interactive inverters, which are used in grid-connected PV systems.

Battery inverters

Many people are familiar with using an inverter when operating standard AC appliances from DC power sources, such as a car battery or larger storage batteries. The inverter takes the DC battery power and converts it to AC for use in AC circuits and loads. These inverters are usually referred to as battery inverters or power inverters. Battery inverters have a wide range of applications and are used in stand-alone PV systems; they are very different from the type of inverter used in grid-connected systems (see Chapter 1). Normally these inverters range from 1kW to 5kW continuous output.

There are also inverter-charger products that are used in off-grid and back-up power systems and these are able to be connected to the grid and can function in a number of different ways: (1) can feed the PV power to charge batteries; (2) can feed power via the inverter to AC loads or to the grid; (3) can feed power from the batteries through the inverter to supply AC power; etc. The various operating options for these inverter-chargers are determined by the individual product and the specifications should be checked closely to confirm this type of product's suitability.

Grid-interactive inverters

This type of inverter, also known as a grid-tied inverter, is used in grid-connected systems and many different models are currently available. They receive the DC input from the array and match it to the AC output required by the utility grid. The inverter will only function when the grid is present and is working within a specific voltage and frequency range. Whether an inverter-charger feeds power back to the grid will be determined by the operating specifications of the unit itself, e.g. a US manufacturer of inverter-chargers has one model which is used only with mains or DC power input and cannot export to the grid and another model which has a software change to allow the inverted AC power to be exported to the grid.

All grid interactive inverters perform these basic functions:

- Convert the DC power from the PV array into AC power that can be used by appliances on site or fed back into the grid via the meter. Without a grid-interactive inverter it is impossible to export the electricity produced by a PV system into the grid.
- Ensure that the power being fed into the grid is at the appropriate frequency and voltage. If the inverter is unable to convert the DC power to match the grid's appropriate frequency and voltage, it will not release electricity to the grid.
- Use 'maximum power point tracking' (MPPT) to ensure that the inverter is finding the maximum power available from the PV array to convert to AC.
- The inverter has inbuilt active and passive safety protections to ensure that the inverter shuts itself down when the grid is not operating within acceptable voltage and frequency tolerances. This is discussed here in the section on inverter protection systems.

Grid interactive inverters may be different in a number of ways depending on:

- whether or not the inverter has a transformer;
- the switching frequency of the transformer used;
- how the PV array and inverter interface with each other;
- the inverter's rated capacity;
- whether the inverter has a single string or multiple string PV power inputs;
- whether the inverter is designed for single phase or multiple phase power supplies.

Transformers

Transformers are devices that use magnetic fields to increase or decrease the voltage of a power source. Traditionally all grid-interactive inverters incorporate

Figure 5.3 An electrician tests a grid-interactive inverter

Source: Global Sustainable Energy Solutions

transformers of varying types, e.g. high- or low-frequency switching. As electronics have improved rapidly, many transformer-less inverter models are now available. Although not widely used in the US, they are becoming popular in Europe and Australia. Transformer-less inverters offer the advantages of being lighter and having a higher efficiency than the traditional product. A transformer, however, can provide electrical isolation because the circuits are connected by a magnetic field (often referred to as 'galvanic isolation') rather than by physical wires as in transformer-less inverters. The disadvantage of transformer-less inverters is that they may inject a small amount of DC power into the grid due to the lack of electrical isolation. Sometimes a small isolating transformer is used to prevent DC injection.

Transformers used in inverters are either low frequency or high frequency. High-frequency transformers are more efficient and lighter than low-frequency transformers but are also more complicated to manufacture.

Figure 5.4 A transformer-less inverter is generally smaller, lighter and more efficient than an inverter of the same size (in kW) with a transformer

Source: SMA Solar Technology AG

Figure 5.5 Inverters with transformers are still the dominant technology in many locations; in the US they are used in almost all PV systems

Source: SMA Solar Technology AG

Margin notes (left column):
- The maximum array power output the inverter can receive
- Array voltage must not exceed this value
- Array current must not exceed this value
- Inverter efficiency
- Physical details
- Inverter must be installed in this temperature range
- Indicates what environment the inverter can be installed in
- Warranty

Technical data	Sunny Boy 3000TL	Sunny Boy 4000TL	Sunny Boy 4000TL/V	Sunny Boy 5000TL
Input (DC)				
Max. DC power (@ cos φ =1)	3200 W	4200 W	4200 W	5300 W
Max. DC voltage	550 V	550 V	550 V	550 V
MPP voltage range	188 - 440 V	175 V - 440 V	175 V - 440 V	175 V - 440 V
DC nominal voltage	400 V	400 V	400 V	400 V
Min. DC voltage / start voltage	125 V / 150 V	125 V / 150 V	125 V / 150 V	125 V / 150 V
Max. input current / per string	17 A / 17 A	2 x 15 A / 15 A	2 x 15 A / 15 A	2 x 15 A / 15 A
Number of MPP trackers / strings per MPP tracker	1 / 2	2 / A: 2, B: 2	2 / A: 2, B: 2	2 / A: 2, B: 2
Output (AC)				
AC nominal power (@ 230 V, 50 Hz)	3000 W	4000 W	3680 W	4600 W
Max. AC apparent power	3000 VA	4000 VA	4000 VA	5000 VA
Nominal AC voltage; range	220, 230, 240 V; 180 - 280 V	220, 230, 240 V; 180 - 280 V	220, 230, 240 V; 180 - 280 V	220, 230, 240 V; 180 - 280 V
AC grid frequency; range	50, 60 Hz; ± 5 Hz	50, 60 Hz; ± 5 Hz	50, 60 Hz; ± 5 Hz	50, 60 Hz; ± 5 Hz
Max. output current	16 A	22 A	22 A	22 A
Power factor (cos φ)	1	1	1	1
Phase conductors / connection phases	1 / 1	1 / 1	1 / 1	1 / 1
Efficiency				
Max. efficiency / Euro-eta	97.0 % / 96.3 %	97.0 % / 96.4 %	97.0 % / 96.4 %	97.0 % / 96.5 %
Protection devices				
DC reverse-polarity protection	●	●	●	●
ESS switch-disconnector	●	●	●	●
AC short circuit protection	●	●	●	●
Ground fault monitoring	●	●	●	●
Grid monitoring (SMA Grid Guard)	●	●	●	●
Galvanically isolated / all-pole sensitive fault current monitoring unit	–/●	–/●	–/●	–/●
Protection class / overvoltage category	I / III	I / III	I / III	I / III
General data				
Dimensions (W / H / D) in mm	470 / 445 / 180	470 / 445 / 180	470 / 445 / 180	470 / 445 / 180
Weight	22 kg	25 kg	25 kg	25 kg
Operating temperature range	-25 °C ... +60 °C	-25 °C ... +60 °C	-25 °C ... +60 °C	-25 °C ... +60 °C
Noise emission (typical)	≤ 25 dB(A)	≤ 29 dB(A)	≤ 29 dB(A)	≤ 29 dB(A)
Internal consumption: (night)	< 0.5 W	< 0.5 W	< 0.5 W	< 0.5 W
Topology	transformerless	transformerless	transformerless	transformerless
Cooling concept	Convection	OptiCool	OptiCool	OptiCool
Electronics protection rating / connection area (as per IEC 60529)	IP65 / IP54	IP65 / IP54	IP65 / IP54	IP65 / IP54
Climatic category (per IEC 60721-3-4)	4K4H	4K4H	4K4H	4K4H
Features				
DC connection: SUNCLIX	●	●	●	●
AC connection: screw terminal / plug connector / spring-type terminal	–/–/●	–/–/●	–/–/●	–/–/●
Display: text line / graphic	–/●	–/●	–/●	–/●
Interfaces: RS485 / Bluetooth	o/●	o/●	o/●	o/●
Warranty: 5 / 10 / 15 / 20 / 25 years	●/o/o/o/o	●/o/o/o/o	●/o/o/o/o	●/o/o/o/o
Certificates and permits (more available on request)	CE, VDE 0126-1-1, DK 5940, RD 1663, G83/1-1, PPC, A54777, EN 50438*, C10/C11, PPDS			
Type designation	SB 3000TL-20	SB 4000TL-20	SB 4000TL-20/V 0159	SB 5000TL-20

* Does not apply to all national deviations of EN 50438
● Standard features   O Optional features   – not available
Data at nominal conditions

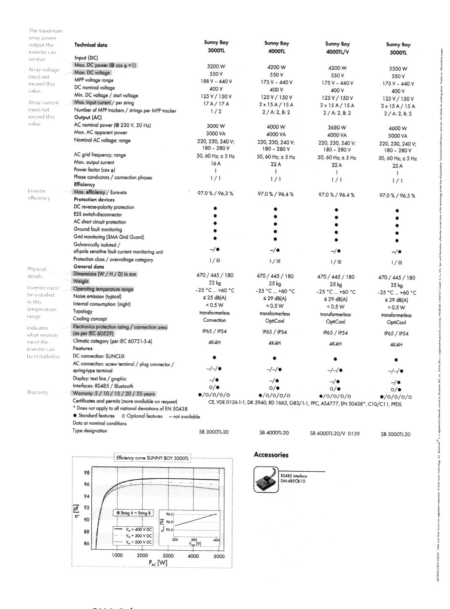

Efficiency curve SUNNY BOY 5000TL

Accessories
RS485 interface DM-485CB-10

www.SMA-Solar.com

SMA Solar Technology AG

Figure 5.6 Just like PV modules all inverters come with a data sheet outlining important information; see also Chapter 9

Source: SMA Solar Technology AG

Mainstream inverter technologies

Here are the general categories of grid-interactive inverters. Within each category there are numerous inverter brands and models available in varying sizes and features. It is very important to use the correct inverter, as discussed in Chapter 9.

Box 5.1 Maximum power point trackers

The maximum power point tracker (MPPT) uses electronics so that the PV array operates to produce the maximum amount of power possible. The MPPT is not a mechanical tracking system, but an electronics-based system that can vary the electrical operating point of the modules to ensure optimum performance and therefore maximum output.

The MPPT tracks the maximum power point (P_{mp}) of the array at regular intervals throughout the day (i.e. because of differing irradiance conditions or shading). The sophisticated electronics of the inverter then convert the maximum power from the array into 220V AC.

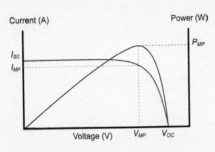

Figure 5.7 I-V curve with the power curve superimposed to find I_{mp} and V_{mp}

Source: Global Sustainable Energy Solutions

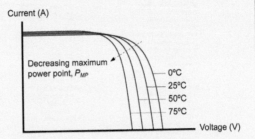

Figure 5.8 The maximum power point falling as temperature increases

Source: Global Sustainable Energy Solutions

String inverters

String inverters are used in small systems ranging from 1kWp to 11kWp. String inverters will all have one maximum power point tracker (MPPT) and the DC input voltages could vary from extra low voltage (ELV) right up 1000 volts DC (low voltage, LV). String inverters can be connected in a variety of ways as shown in Figures 5.9, 5.10 and 5.11.

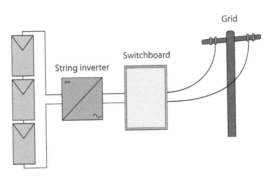

Figure 5.9 String inverter connected across one string of PV modules. The PV array comprises two parallel strings with a single input to the inverter, which means the inverter is only receiving power from the one combined PV string; this is a more stable set-up as there is no potential for interference from other strings

Source: Global Sustainable Energy Solutions

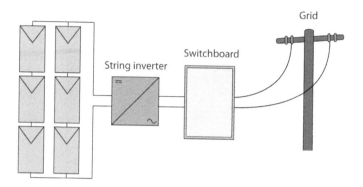

Figure 5.10 It is possible to connect multiple strings across the same string inverter, but if one string has a lower output (due to shading or damage) it will affect the output of the entire array

Source: Global Sustainable Energy Solutions

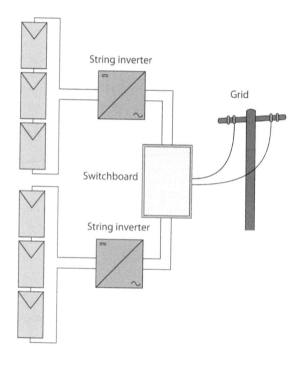

Figure 5.11 Multiple inverters can be used to increase the reliability of the system; if one inverter malfunctions, the others will still work and the system can continue to export power

Source: Global Sustainable Energy Solutions

Multi-string inverter

A multi-string inverter is a single inverter appliance, but it has a number of MPPT inputs. Therefore the PV array can be divided into multiple strings and a suitable combination of strings connects to each one of the inverter's MPPT inputs.

These inverters have the advantage that if the modules are facing different directions then the array could be divided into strings so that modules in the same string are all facing the same direction. These individual strings then connect to a dedicated MPPT so that the energy yield from the system will be

Figure 5.12 Multi-string inverters installed by Solgen Energy on Cockatoo Island for Sydney Harbour Federation Trust

Source: Solgen Energy

greater than if the strings were connected to an inverter with only one MPPT. A multi-string inverter is generally cheaper than using a number of individual inverters and can offer the advantage of higher energy output for arrays where parts of the array face different directions or experience different levels of shading.

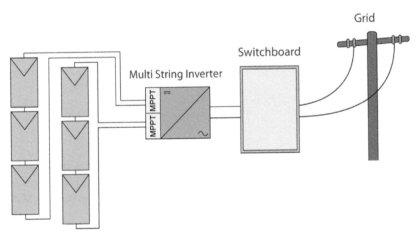

Figure 5.13 Two strings are each connected to different MPPTs so that if one is shaded it will not reduce the output of the other

Source: Global Sustainable Energy Solutions

Central inverter

A central inverter is very similar to the string inverter with multiple strings – the difference is that central inverters are generally used for a large system (>10kWp). In these systems the array could be divided into a number of sub-arrays, each comprising a number of strings.

In some systems there is just one large inverter suitable for the whole array or the single central inverter might be a single enclosure housing several smaller multi-string inverters combining to represent a single, electrical output. In others there will be a number of inverters, for example 5 × 20kW inverters for a 100kW system.

Some manufacturers of these large central inverters have rationalized their design: e.g. a central inverter can comprise multiples of smaller inverters that can operate selectively depending on the amount of power that can be produced from the available sunlight. This configuration improves the operating efficiency of the central inverter unit, especially when required to operate at less than peak load.

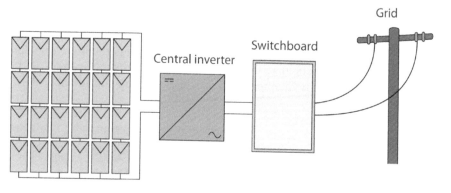

Figure 5.14 Connecting multiple strings to a central inverter

Source: Global Sustainable Energy Solutions

Figure 5.15 For very large installations a whole room of central inverters may be required. These are SMA Sunny Mini Central inverters

Source: SMA Solar Technology AG

Modular inverters

Modular inverters (also known as micro-inverters) are small transformer-less inverters (some will have an isolating transformer to minimize DC injection currents) designed to be mounted on the back of the PV module.

Over the years there have been a number of modular inverters manufactured in the range 100–300W. This product has re-emerged during 2009 for the grid-connect PV market. Two main advantages of the modular inverter are that they remove the requirement for DC cabling from the array as each module has an AC output and these AC cables can be paralleled at each module and then connected to the grid at the appropriate location. These inverters are also small and easy to handle, and they have the advantage of being modular (just like PV modules), which means that more modules and inverters can be added to the system in future at minimum cost.

In the past, modular inverters were generally more expensive, i.e. the cost per unit of power ($ per W) was higher than other inverters; but the latest modular inverters on the market are a similar cost $/watt to installing a single inverter for a PV installation.

The disadvantages of modular inverters are related to the fact that they are installed on the back of PV modules. If the inverter fails, repairing or replacing it involves removing the modules from the array to access the inverter behind it. In addition, designers may be wary of using modular inverters in hot climates. As discussed in Chapter 4, PV modules heat up considerably during

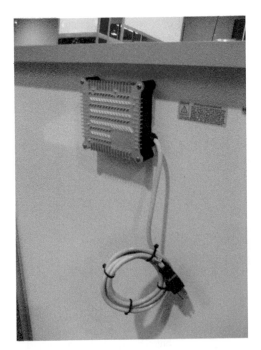

Figure 5.16 Micro-inverter attached to the back of a module

Source: Global Sustainable Energy Solutions

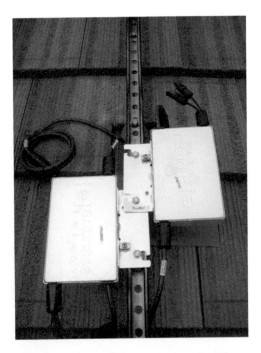

Figure 5.17 Micro-inverter available on the US market

Source: BMC Solar

the day, so the inverter will operate at a higher temperature than it would be if it were located on a shaded wall or indoors, increasing the risk of inverter failure. The operating characteristics of modular inverters under high temperatures should be provided by the manufacturer.

These latest micro-inverters offer similar data logging and communications features as other inverters on the market either via the AC power system at the location or via a website.

Table 5.1 Inverter types and characteristics

Inverter type	Modular	String	Multi-string	Central
Power range	100–300W	700–11,000W	2000–17,000W	10,000–300,000W
MPPT	Yes	Yes	Multiple	Multiple
Typical efficiency	95%	93–97%	97%	97%
Advantages	No DC cabling; easy to add more modules	Readily available	Multiple MPPTs; readily available	Lower $/W cost; one location for maintenance
Disadvantages	Replacing a faulty inverter can be difficult	Only one MPPT		No redundancy if inverter fails

Inverter protection systems

Grid-connected inverters will only work if the AC grid is functioning and is within the grid's predetermined operating conditions. If these conditions are not met, the grid-connected inverter will disconnect and will not output any power from the PV array at all. The inverter is set up to mirror the function of the grid itself. The MPPT software in these inverters allows the PV output to be optimized to best match the grid specifications at the time of power output. Grid-interactive inverters will typically incorporate two types of protection: active and passive. Both forms have the inverter switch off on over/under frequency or over/under voltage. This protection is intended as self-protection for the inverter if extreme conditions occur and protection for the grid itself, so the inverter will disconnect if it cannot see the grid, e.g. if there is a blackout.

Self-protection

Inverters incorporate protection mechanisms for a variety of problems:

* Incorrect connection: if an inverter is incorrectly connected to the PV array (e.g. with reverse polarity) it will not work and will in most cases be damaged. Even though some inverters protect against incorrect connection, the warranty for most inverters does not cover such damage.
* Temperature: inverters are sensitive to temperature variation and manufacturers will specify a temperature range within which they must operate. Some inverters reduce their power output or turn themselves off when temperature increases past the manufacturer's operating specification. Although the inverter might have over-temperature protection, it is important that the inverter has sufficient ventilation and cooling; over-temperature can cause damage to the inverter.

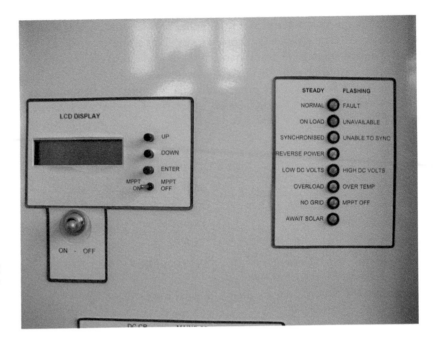

Figure 5.18 The front panel of an inverter will commonly display array faults

Source: Global Sustainable Energy Solutions

- DC voltage too high: all grid-connected inverters will have a specified voltage range within which they operate correctly. Some inverters switch off to protect their electronics if the maximum DC voltage they can tolerate is exceeded, but the inverter could still be damaged; other inverters have no such protection.

Grid protection

Grid-interactive inverters must be able to disconnect from the grid if the supply from the grid is disrupted or the grid itself is operating outside the preset parameters (e.g. under/over voltage, under/over frequency). In both these cases, the inverter disconnects to avoid continuing to output power to the grid when the grid itself is not operating.

There has always been concern that, if there were a sufficient number of inverters connected to the grid in an area and the grid supply to that area failed (e.g. a car hitting pole and cables breaking), the inverters would interact with each other, meaning that the voltage and frequency will become a reference to each other (i.e. an inverter would still think the grid was 'on' and continue exporting electricity to the grid). In this circumstance the passive protection, which switches the inverter off when the grid is operating outside the required voltage and frequency, may not operate, and power would continue to be exported onto a grid that is not operating. This phenomenon is known as 'islanding'. Therefore active as well as passive protection is required. 'Islanding' is a serious safety concern for electricity utilities, e.g. if the grid shuts down and technicians are working on it, there must be certainty that all grid-connected PV systems are also disconnected from the grid. Islanding obviously poses a significant risk of electrocution to workers trying to repair power lines and may damage transmission equipment.

The 'islanding' issue is met by the inverter's active and passive protection. Many standards require grid-interactive inverters to have both these features. Passive protection is provided by the inverter's ability to detect the grid's voltage and frequency, i.e. if the inverter detects that the grid is either over or under voltage and over or under frequency, it will shut down. Active protection is provided by the inverter detecting any frequency instability, frequency shift or power variation that would vary the voltage that the inverter detects. The inverter will shut down via this active protection if it detects any of these conditions.

When the condition that caused the protection device to activate is removed and the inverter synchronizes again with the grid, the inverter will reconnect to the grid after a period of time, e.g. in general at least one minute. Some countries require inverters sold in their territory to have a minimum time delay between the inverter detecting a stable grid and the inverter reconnecting. This time delay is generally programmable within the inverter.

Balance of system equipment: System equipment excluding the PV array and inverter

In addition to the PV array and inverter, a system requires a variety of other components in order to function. These are known collectively as the balance of system equipment (BoS) and often must comply with local and/or national codes and regulations. The BoS equipment is composed of the components required to connect and protect the PV array and the inverter. This equipment includes cabling, disconnects/isolators, protection devices and monitoring equipment. The key balance of systems components are given below, and are described in further detail:

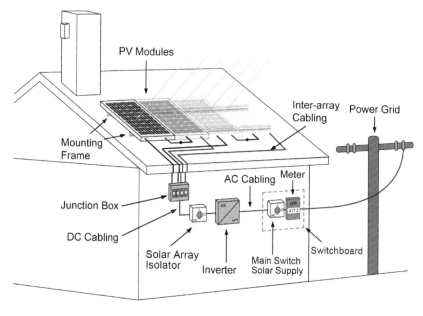

Figure 5.19 Key balance of system components shown excluding grounding/ earthing

Source: Global Sustainable Energy Solutions

- DC cabling including the inter-array cabling (cabling used to connect the various modules and strings together to form a PV array), the cable from the PV array to the PV combiner box (if required) and the cable from the PV combiner box to the inverter.
- PV combiner box, normally only required when the PV array has more than one parallel string (see Chapter 4) and is located between the PV array and inverter.
- Module junction boxes located at the back of each module; they normally bring together the wires used to connect the PV cells to form a module.
- Protection and disconnect devices, such as the DC and AC main disconnects/ isolators, are often required by many local standards and codes.
- Lightning and surge protection.
- Metering: the building will already have an electricity meter used to measure electricity flows in and out of the building. Installers may incorporate this meter into the new PV system or install a new meter, depending on whether the current meter meets the requirements of the system. Meters are either gross or net; the differences are important and are discussed below.
- AC cabling connecting the inverter to the meter, and the meter to the utility grid.
- Grounding/earthing cables for the array.
- Monitoring: most PV systems incorporate some kind of monitoring so that the owner is able to see the outputs of their system and any problems, such as a decrease in power output, can be quickly identified.

Cabling

During the 1990s, with increasing interest in grid-connected PV systems, ways to the make the installation time shorter and easier were investigated. A number of manufacturers developed plug and socket connections that are now very commonly used.

These have a positive and negative plug which minimize the risk of incorrectly interconnecting modules in series. They will typically have a locking mechanism so that they are not easily disconnected by pulling on the cable.

Many module manufacturers now supply their modules with the cables and plugs already connected into the junction box (see below) on the back of the

Figure 5.20 Example of solar cable plugs

Source: Multicontact

Figure 5.21 Another example of solar cable plugs

Source: Multicontact

PV module. The cable and the plugs can also be individually purchased so that the cables can be made to a length to suit the actual installation; longer cables are often required for certain installations.

It is important to select a cable that meets the output current and output voltage of the PV array and minimizes voltage drop. Typical cable sizes available include: $2.5mm^2$, $4mm^2$ and $6mm^2$. These sizes are suitable for the majority of installations. Selecting cable sizes is discussed in Chapter 10. The AC and earthing cables used for small-scale grid-connected photovoltaics are the same as those typically used in buildings.

PV combiner box

If the array is composed of a number of parallel strings, then the cables from the array strings may be interconnected in a PV combiner box, sometimes called an array junction box or DC combiner box. Even if the array has only one string, PV combiner boxes can be used to interconnect the output cables from the array to the array cable to the inverter, particularly if the array cables are larger in diameter than those used for interconnecting the actual modules within the array.

If there are multiple parallel strings then the PV combiner box will facilitate the combining (connection) of the positive and negative cables from the different strings on links (or similar) allowing only one positive and negative array cable to interconnect with the inverter (via the DC main disconnect/isolator).

Figure 5.22 PV combiner box with over-current protection devices, in this case circuit breakers. Mounting them in the PV combiner box provides easy access

Source: Global Sustainable Energy Solutions

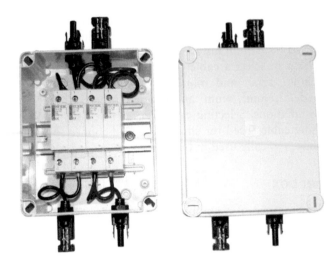

Figure 5.23 PV combiner box showing string fusing provision

Source: DKSH Australia

Module junction box

In some standard PV modules the wiring connections are enclosed in a module junction box attached to the back of the modules. These boxes include knock-outs where either cable or conduit glands can be installed. Cables can then be directly installed into the terminals of the junction box. In some PV modules, the module junction box is permanently sealed in the back.

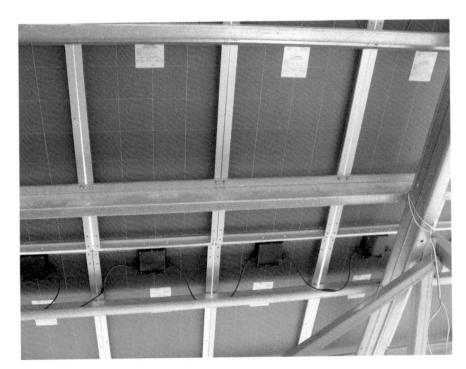

Figure 5.24 Module junction boxes on the back of modules in a PV array

Source: Global Sustainable Energy Solutions

Circuit breakers and fuses

Fuses and circuit breakers are commonly used as over-current protection in PV systems. National codes should specify over-current protection requirements. Normally either DC fuses or DC circuit breakers are used on the array side of the inverter and AC fuses or AC circuit breakers on the grid side. DC fuses may be required by codes in PV systems and they are very different from the AC fuses used in regular appliances. It is much more difficult to break a direct current, so it is important to make sure that any fuses used for DC applications are DC-rated. Most DC-rated fuses can be used in AC applications (manufacturers will generally state this), but DC fuses are more expensive. Likewise, AC circuit breakers are not compatible with DC circuits and vice versa. It is also important to note that DC fuses and DC circuit breakers in a PV system operate very differently to AC fuses and AC circuit breakers in AC systems, because PV is a current-limited source (i.e. the maximum current an array will produce is its short-circuit current I_{sc}). In an AC system, fuses and circuit breakers will blow/open very quickly because the current produced under fault conditions is very large, whereas in PV systems the current produced under fault conditions (I_{sc}) is not much higher than the normal operating current (I_{mp}) and so may not be interrupted by the string fuses and circuit breakers. It is imperative that those installing a PV system are qualified to work with electrical systems (specifically PV systems) and are familiar with local codes governing the choice of over-current protection and its installation.

A fuse is a device fitted to protect against excessive current flows that could damage conductors in a circuit, and to reduce the risk of fire due to overheating of conductors. This will commonly consist of a short section of conductor of sufficient size to carry the load current mounted in an insulating enclosure; under fault conditions the fuse will open the circuit. Fuses may be either rewireable or a cartridge high rupture capacity (HRC) style. Rewireable fuses are no longer considered sufficient to protect a wiring system, so cartridge fuses should be used at all times. Fuses normally feature a current rating, i.e. the amount of current that can pass through the fuse before it melts. Circuit breakers are mechanical devices that will open the circuit under fault conditions and the switch can be manually flipped to close the circuit and restore current flow when the fault has been removed. Currents in excess of the fuse or circuit breaker rating will cause the device to operate (open) and prevent any current from flowing.

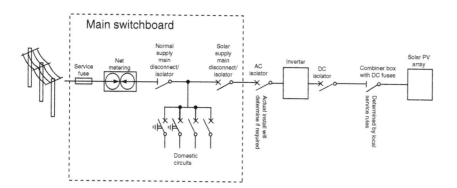

Figure 5.25 When considering possible array wiring, always check that potential designs conform to local wiring codes/standards

Source: Global Sustainable Energy Solutions

Figure 5.26 Circuit breakers in a grid-tied PV system

Source: Global Sustainable Energy Solutions

Fuses and circuit breakers may be used, not only as fault (excess)-current protection, but also as disconnect devices, as explained below. Fault-current protection is found on the DC side of the system (arrays, sub-arrays and strings), as well as on the AC side of the system (there is usually a circuit breaker on the utility side of a PV installation). Over-current protection requirements and ratings vary by country so it is important to check national codes.

PV main disconnects/isolators

A disconnect/isolator isolates equipment from electricity and power sources and allows the power in a circuit to be shut down. Disconnects/isolators should be installed on both the DC (typically DC circuit breakers are used) and AC sides of the inverter as shown in Figures 5.27, 5.28 and 5.29.

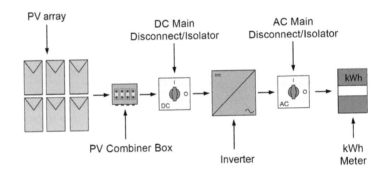

Figure 5.27 Locations of DC and AC disconnects/isolators

Source: Global Sustainable Energy Solutions

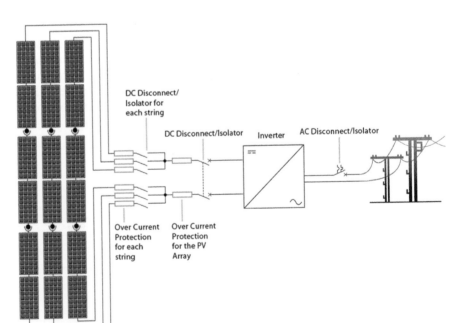

Figure 5.28 Diagram of a three-string PV system showing protection devices including AC and DC disconnects/isolators and over-current protection

Source: Global Sustainable Energy Solutions

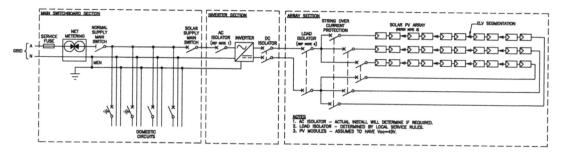

Figure 5.29 Schematic showing main disconnect/isolators and over-current protection. ELV segmentation is explained in Chapter 8

Source: Global Sustainable Energy Solutions

Box 5.2 Utility external disconnect switch (UEDS)

Inverters sold in the US normally have a utility external disconnect switch (UEDS) as a means to disconnect the system on the AC side of the inverter from the grid when required, in addition to the AC circuit breaker in the switchboard. The UEDS is typically incorporated into or installed near the inverter, which is in a location easily accessible to utility personnel should they need to switch off the system for routine or emergency maintenance on the local grid. UEDSs were previously mandatory, but are no longer required by all utilities and states for small systems; they are not widely used outside the US. Local and national codes and regulations should be checked to determine if UEDSs are recognized as an approved AC isolator by utilities and regulators. Remember a properly functioning inverter will not operate when the grid is down, which is why many solar professionals consider these switches to be largely redundant.

Figure 5.30 These inverters are designed for the US market and incorporate a large UEDS beneath the inverter

Source: Global Sustainable Energy Solutions

Lightning and surge protection

Lightning and surge protection may or may not be required, depending on the system and local codes. If it is required then the devices may be required on both the DC side of the inverter (protecting from strikes on the array) and on the AC side of the inverter (protecting from strikes on the AC power grid).

The exact positioning of these protection devices must be in accordance with the protection device manufacturers' recommendations. It is common for

Figure 5.31 These miniature circuit breakers (MCBs) provide over-current protection for the inverter

Source: Global Sustainable Energy Solutions

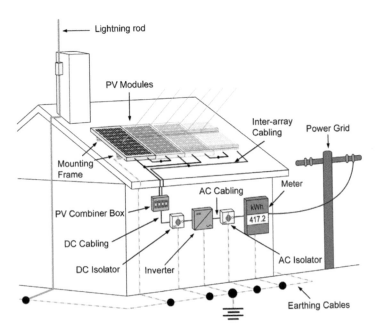

Figure 5.32 System installation layout showing lightning protection and common grounding/earthing point as per Australian national codes; methods may vary by location and the national codes should always be followed

Source: Global Sustainable Energy Solutions

them to be located in the PV combiner box as the array wires are already required to be close to each other. In addition, if remote monitoring of the inverter is possible through a modem, then protection devices should be connected to the telephone line on the line side of the modem.

System monitoring

PV installations may use a data logger to measure and record information about the performance of a system. The information recorded may include the time, power, voltage and temperature of a system. This information might then be sent to a central website hosted by the inverter manufacturer, be displayed on a school or company website, or on a display somewhere.

Metering

An electricity meter records the electrical energy in kWh consumed by the loads within the building where the meter is connected. The meter records the number of units of energy consumed and a unit is typically 1kWh. The electricity consumer is then charged for this electricity based on the price set for that consumer. Electricity distributors will often have different rates for residential homes, industrial and/or commercial consumers.

Figure 5.33 Example of a software data logger

Source: Global Sustainable Energy Solutions

Figure 5.34 Rotating barrel style meter, here shown accompanied by signage for a PV installation

Source: Global Sustainable Energy Solutions

Figure 5.35 A more complex digital meter also known as a smart meter; many governments and utilities have started to introduce smart meters, which are expected to be very widely used in future

Figure 5.36 The single rotating disc in the centre of this net meter is clearly visible: when electricity is consumed (by loads within the building) it rotates forwards and when electricity is produced (by the PV system) it rotates backwards

Figure 5.37 An electricity meter commonly used in the US

The simplest meter is a mechanical device with a calibrated rotating disc that spins when electricity is being consumed, as shown in Figure 5.36. A more complicated digital meter like the one in Figure 5.35 can record the time of day that the energy is consumed. This type of meter is used when electricity prices vary according to the time of day. The type of meter installed with a grid-connected PV system will depend on the electricity provider and the region.

Net metering

Net metering is a method by which a utility measures the difference between the consumption of a site and the generation at the site. In a typical residential system the electricity produced by the system will be exported to the grid during peak sunlight hours (usually 10.00am to 3.00 or 4.00pm) and the consumer will import electricity from the grid for use in the evening. If the generated energy is less than the consumed energy, then there is no net export and the customer pays the utility for the difference. If the generated energy is more than the consumed energy there is a net excess generation (NEG), the utility may pay the customer for their NEG or roll it over to offset the next month's bill. Net metering is widely used and is very common in the US where many states' utilities must make net metering available to customers with grid-connected PV systems.

The simplest metering method to achieve the net-metering effect is to allow the mechanical meter to operate in both directions. In this arrangement, electricity produced by the PV system either provides power directly to the loads or is exported to the grid, making the meter rotate backwards, reducing the actual number of units consumed as counted by the meter. In the evening, as the electricity for the loads is provided by the grid, the meter will rotate forwards, thereby increasing the number of units consumed as counted by the meter. In this arrangement the meter is effectively a net import meter and the customer is only charged for the units that are imported and thereby recorded on the meter.

The disadvantage of this metering arrangement is that it neither informs the user of the exact quantity (in kWh) that the PV system has produced nor the exact quantity that they have consumed. There is no record of the amount of electricity that is supplied directly from the PV system to the loads within the building.

It is recommended that if the local distributor requires this metering arrangement, then an installer should install a separate meter (if not included with inverter) that records the exact quantity of electricity produced by the PV system. This will allow analysis of system performance and comparison with the figures on the export meter (if used) to determine how much electricity has supplied loads directly within the building.

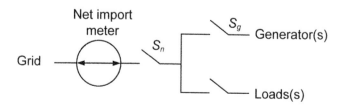

Figure 5.38 Diagram of a net import meter able to run in both directions

Source: Global Sustainable Energy Solutions

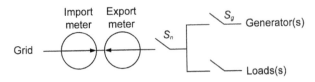

Figure 5.39 Diagram of a net import meter using two meters – one to measure import and the other to measure export. The NEG can be calculated simply by subtracting the number on the export meter from the number on the import meter

Source: Global Sustainable Energy Solutions

Gross metering

Gross metering is a method used by utilities to measure the entire solar energy production and electricity consumption separately for a site. This method is common in places such as Europe and Australia where gross feed-in tariffs are available (see Chapter 13). Gross meters either have two spinning discs (one for consumption and one for production) or two mechanical meters installed, which only operate (or rotate) in one direction. The export meter records the amount of electricity generated by the PV system exported to the grid during the day, while the import meter records the exact amount of electricity consumed from the grid.

In this arrangement the amount exported can be deducted from the import meter and the customer will be charged for the net imported (net metering) or exports and imports may be differentially priced, allowing the customer to make a profit. Gross metering can also be achieved by using a dual electronic import and export meter.

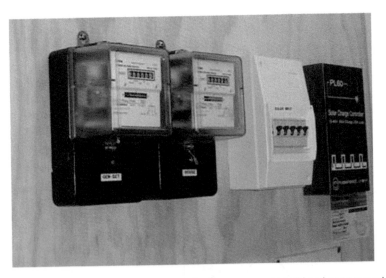

Figure 5.40 Two metering devices, one measuring the power generated and one measuring the power used

Source: Global Sustainable Energy Solutions

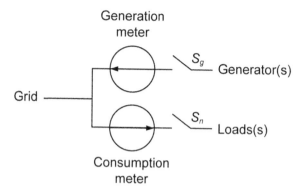

Figure 5.41 Diagram of a gross-metering system

Source: Global Sustainable Energy Solutions

Figure 5.42 Multiple meters are common, often for off and on peak demand metering and large appliances like ducted air-conditioning

Source: Global Sustainable Energy Solutions

6
Mounting Systems

This chapter discusses different techniques used to mount PV arrays. First, roof mounting systems are explored; these are commonly used in small-scale urban installations and four different techniques are covered. Second, two different techniques for ground mounting systems are explained; ground-mounted systems are generally used for very large installations but may also be used for small installations on large properties where space is plentiful, such as farms. sun-tracking systems that turn PV modules to follow the path of the sun and capture a larger amount of solar radiation are also explained. The chapter concludes with a discussion of the safety issues surrounding mounting systems.

It is important that installers familiarize themselves with the local requirements for the installation of a PV array. Some regions require that an approval process be met before the installation is allowed. It is important that the manufacturer's installation guidelines are followed so that the product warranty is not compromised and any structural certification remains valid. Always ensure that the mounting system is compatible with the PV modules to be installed.

Roof mounting systems

For homes or businesses using a grid-connected PV array, the most common installation is the rooftop mounting system. Its most important role is to securely and safely attach the solar array to the roof. Aside from safety there are three other important factors to consider when choosing a roof mounting system: the amount of solar radiation the module will receive in that position, ventilation of the module and the overall aesthetics of the PV system.

The amount of solar radiation the module receives (the module's solar access) will directly affect the power it produces and therefore should be optimized by using a mounting system that secures the array at optimal orientation and tilt angle for that location. In cases where attaching modules at the angle and direction of the roof will not yield this result, installers may consider using a mounting system that can elevate the modules to face the optimum tilt angle and orientation and so improve the modules' power output.

Ideally a mounting system should allow for as much clearance as possible above the roof in order to increase the flow of air around the modules. This ventilation provides convective cooling and can reduce the temperature the modules heat up to during the day. As discussed in Chapter 3, modules perform better at cooler temperatures and ventilation can improve the performance of the array, while not allowing any ventilation behind the modules is likely to decrease the modules' output.

Figure 6.1 This mounting scheme allows good ventilation. Mounting panels closer to the roof increases operating temperature and thereby lowers system performance

Source: Global Sustainable Energy Solutions

Figure 6.2 This mounting scheme will not permit good airflow

Source: Global Sustainable Energy Solutions

The aesthetics of the PV system is often a priority for homeowners. Systems discreetly installed close to the roof, blending with the architecture, are often considered more aesthetically pleasing. This visual appeal comes at a cost, particularly to ventilation. When modules are installed close to the roof, heat will have less chance of escaping from underneath the array than if the modules were elevated above the roof. Thus, low-profile mounting means the PV modules will probably operate at higher temperatures, with reduced power output as a result. Additionally a system installed close to the roof will follow the tilt angle and orientation of the roof; if these are not optimal then it will adversely affect the power output of the PV array.

The balance between these trade-offs is complex and the installer should provide the customer with information regarding the trade-offs and expected performance of the system using each mounting system so that the customer can make an informed decision. The different types of roof mounting systems are outlined below. All these are fixed systems, which do not move and, as explained, each has its own unique benefits.

Pitched roof mounts

Pitched roof mounts are the most common roof-mounted system because they are versatile, easy to install and relatively inexpensive. These systems are typically mounted just above the roof surface at the same orientation and tilt angle as the roof. Pitched roof mounts are normally attached to the roof's structural members, e.g. the rafters, through the use of lag bolts or fixing brackets.

A horizontal railing system is then attached to these brackets and modules are secured to the railing through the use of module clamps. In this way the railing elevates the modules above the roof surface and allows for increased air circulation under the array. Pitched roof mounts are popular because they allow for some ventilation around the module while still keeping a fairly low profile and are thus considered an excellent balance between aesthetics and ventilation. Commonly used materials in pitched roof mounting systems include aluminium, stainless and galvanized steel.

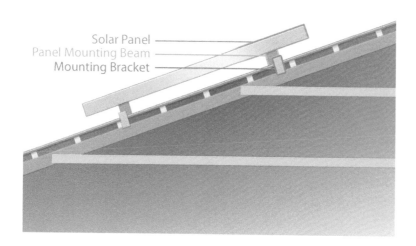

Figure 6.3 Pitched roof or standoff mounts are normally attached to rafters

Source: Global Sustainable Energy Solutions

Figure 6.4 Example of a pitched roof mounting system clearly showing the module clamp

Source: Global Sustainable Energy Solutions

Pitched roof mounts for tiled roofs

Installing a PV system on a tiled roof can take longer than on a metal roof because of the time-consuming process of removing and replacing tiles around the areas where the mounting system is connected to the roof (attachment points). Installers often use a roof hook that attaches flush to the rafters underneath the tiles. The hook has a goose neck design and a protruding arm for the attachment of a rail mounting system. Different manufactures have varying designs to connect to the top or the side of the rafter, designed for different tile types.

Figure 6.5 Tile hook for lateral rail connection

Source: Global Sustainable Energy Solutions

When planning an installation on a tiled roof it is also important to consider tile replacement. Tiles are easily broken by installers walking on the roof and any cracked or damaged tiles must be repaired or replaced. It is good practice to find out about the availability of spare roof tiles before beginning an installation.

Pitched roof mounts for metal roofs

Working on a metal roof poses safety risks due to the conductivity of metal. Care should always be taken when working on a metal roof that the bare ends of any active cables do not come into contact with the roof. Dissimilar metals must not come into contact as this could cause galvanic corrosion, damage the roofing material and mounting system, and pose a significant safety risk to occupants of the building. Galvanic corrosion can be prevented by inserting a rubber separator between dissimilar metals, a technique known as galvanic isolation (also referred to as galvanic insulation). For this reason, manufacturers usually list approved fasteners with their panels and using these fasteners is advised.

Certain types of metal roofs can also use mounting techniques that do not require the system to penetrate the roof material. These techniques are particularly popular in locations where there is frequent rainfall and the roof must be absolutely watertight.

Be aware of a hot climate's effect on metal roofs. In some areas metal roofs can become very hot, requiring extra ventilation to maintain reasonable panel operating temperatures. Seeking information from local trade associations and experienced industry operators can give insight into potential temperature effects in your area and how best to address them.

Figure 6.6 Corrugated metal roof installation

Source: Global Sustainable Energy Solutions

Rack mounts

Rack mounts (also referred to as tilt-ups) are used when the roof slope or orientation is far from the best; they are most commonly used on flat or low-angle roofs. Rack mounts utilize most of the same mounting hardware as pitched roof mounts, but differ chiefly in that they include a triangular-shaped support structure to elevate the array and increase its tilt angle. If installed on a flat roof, there is often no need for scaffolding and in some circumstances no need for a harness (check local working-at-heights legislation). Despite ease of installation, this mounting option typically involves extra engineering costs because of the additional weight of the structure on the roof as well as the greater exposure to wind loading forces (discussed later in this chapter). An installer should consult with their local authority or council to find the local wind speed design requirement as well as any other special design cases that must be considered before installing a rack-mounted solar array. Usually such information can be found in the relevant local codes/regulations. Determining which codes/regulations apply can be quite difficult and consulting local industry operators can help.

In most locations the rack-mounted system is required to be approved by a structural engineer to ensure it is appropriately designed for wind loading. Before choosing this option these additional costs, calculations and aesthetic impacts should be offset against the value of the increased energy yield (from using the optimum orientation and tilt angle).

Providers of complete array mounting systems often provide a set of installation instructions, which, if followed, will mean the installed array will meet wind loading standards. A certificate/statement to show the installation's compliance should be included with the system logbook to prove its conformity to standards, if required by system inspectors.

Figure 6.7 Raked mounting system

Source: Global Sustainable Energy Solutions

Figure 6.8 This PV system has been designed to discreetly blend in with the architecture through the use of a curved direct mounting system

Source: Global Sustainable Energy Solutions

Direct mounts

A direct mount (also known as flush mount) is a PV mounting method where modules are installed directly on the surface of the roof, so there is very little space between the module and the roof surface. While this mounting method offers superior aesthetic appeal, the lack of ventilation beneath the array is likely to reduce the array's power output when compared to pitched roof or rack mounted system.

Building-integrated systems

Building-integrated photovoltaics (BIPV) are becoming increasingly popular as they are considered more aesthetically pleasing than the traditional roof mounting systems. BIPV has caught on with eco-minded architects because of its ability to be seamlessly integrated into a building, either as a part of the roof, integrated into skylights or windows, or simply mounted directly on the building's facade.

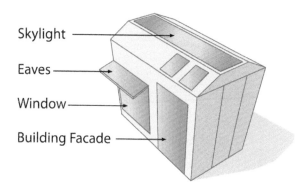

Figure 6.9 Examples of where PV may be integrated into a building

Source: Global Sustainable Energy Solutions

BIPV projects tend to be highly customized and are generally much more complex than standard roof installations. BIPV often requires architects and engineers to work closely together to ensure that a system is safe and waterproof while blending into the architectural features of the building.

Solar tiles and shingles are a form of BIPV which is becoming more common in new home construction because they can be used in place of traditional roof tiles. Unlike traditional PV mounting, BIPV tiles are not elevated above the roof, which creates a more aesthetically pleasing system.

Figure 6.10 BIPV awnings developed by Sunvie and integrated into a car park; this 1.15MWp installation is located in St Aunès, France

Source: Sunvie, www.sunvie.eu

Figure 6.11 Example of solar tiles as part of complete roof

Source: Nu-lok Solar and Barker Electrical

Amorphous laminate modules are also available. These are long, narrow and flexible modules with an adhesive layer on the back surface that can be used to secure the module onto the roof. Amorphous laminates can be used on several different types of wide-pan, metal roofing sheets and other materials such as membrane roofing materials. This form of BIPV is especially good for buildings in areas that are likely to experience very high wind speeds as they pose very little resistance to airflow.

Concluding the discussion on roof-mounted PV systems, it is important to note that some mounting systems may not be compatible with certain solar modules, so it is critical to check the installation instructions provided by the solar module manufacturer before selecting a mounting system. Manufacturers will have prescribed the installation requirements for their PV modules, such as where clamps should be attached and where support rails are needed.

Most mounting systems will use two horizontal rails per row of modules. It is generally recommended that the rails be spaced far enough apart so that 50–74 per cent of the module is between the two rails. It is also important to note that if the rails run parallel to the rafters (down the roof), the space between railings may be determined by the space between the rafters that they must adjoin and it is important to check that this won't prevent the system from being installed as per the manufacturer's recommendations.

Manufacturers also commonly recommend that modules be clamped on their longer side as clamping on the shorter side may not allow for the structural integrity of the module to withstand strong suction forces caused by wind blowing over a module, which can cause buckling of the module and cracks in the glass. Clamps should be used that do not shade any cells in a module and only extend over the module frame without coming into contact with the glass, as this could apply undue force onto the glass, causing the glass to crack.

Figure 6.12 Amorphous laminates installed on metal roofing in Australia

Source: Global Sustainable Energy Solutions

Ground mounting systems

Ground-mounted PV systems have two main applications; they are commonly used on residential or commercial properties where rooftop mounting is not a viable option and there is plenty of free ground space, or for very large-scale PV installations. Ground-mounted systems have many advantages: first the tilt angle and orientation is not constrained by the slope or direction of a roof or facade and so the array can be installed at the optimum tilt angle and orientation; they offer the installer the ease of constructing a system at ground level, and avoiding work at heights. Working without ladders and lifts saves time for an installation in some cases, although ground-mounted systems may also require civil engineering, trenching and storm water management. The additional costs of material, labour and engineering associated with ground-mounted PV arrays might not be economical for many small- or medium-sized grid-connect systems.

Local building laws and planning regulations should be adhered to, and this may include a planning application with the local authority. The system should be situated away from site boundaries and a surveyor may be needed for this. It is also important to avoid underground water services and electrical and telecommunications cables. These can be identified by contacting local water and electrical utilities and acquiring a map showing possible hazards. Many locations also have services such as 'Dial Before You Dig' set up for these purposes and for use by the construction industry.

Depending on local codes, a fence may be required to prevent unauthorized persons from accessing the array. Fencing is generally recommended for high visibility ground-mounted arrays to help discourage theft and vandalism. Large-scale systems may even include security cameras, motion sensors and advanced monitoring equipment to reduce the likelihood of theft.

Ground rack mounts

Similar to roof-rack mounts, ground-rack mounts allow flexibility in orientation and tilt. Ground-rack mounts are a fixed tilt mounting system and often use a

Figure 6.13 Large ground-rack mounted installation

Source: Global Sustainable Energy Solutions

Figure 6.14 The grass is low now, but if it grows high enough it could shade the array

Source: Global Sustainable Energy Solutions

pre-engineered steel or aluminium array frame to securely hold the modules in place. Modules are normally clamped or bolted onto the frame, which is fixed in a concrete foundation. Where the installation site wind conditions and soil are suitable, earth screws or driven piers may be suitable replacements for a concrete foundation.

Large ground-rack mounted PV systems normally require the PV arrays to be installed in rows in order to fit into the given space. A drawback of this method is that arrays may shade the rows behind them and as discussed in Chapter 3 even a little shading on a PV module dramatically reduces its output, especially very early or late in the day when the sun is low in the sky and the arrays cast long shadows. An installer must choose a row spacing to avoid shading throughout most of the year; this can be calculated using trigonometry, or a general rule of thumb for most installation locations outside the tropics is that the minimum distance between rows should be three times the height of the mounting structure. Providers of mounting hardware often provide software for the necessary calculations. Closer to the tropics, it might be possible to decrease the spacing between rows. But since shading is a major concern, a thorough shade analysis should be conducted before installing the system. Ground-rack mounted systems may also have to take into account maintenance requirements, for example, if an array is installed in a field grass-cutting must be considered as an ongoing maintenance cost.

Pole mounts

Pole mounts are popular for systems requiring fewer modules. The main advantage of these systems is that they are an inexpensive option because they do not require many installation materials and they are usually adjustable so the installers can change the tilt angle seasonally to keep the array at the optimum tilt angle for a larger part of the year.

Figure 6.15 Pyramid-style pole mount, which also uses a sun-tracking system

Source: Global Sustainable Energy Solutions

Sun-tracking systems

Sun-tracking systems are mechanisms that turn the array to ensure it is always facing the sun and thereby operate the solar modules at peak power for a longer period of time each day. These systems are much more expensive. However, where space is limited or the value of the increased energy harvest is high, the additional cost of the tracking system may be justified over the life of the project.

Most tracking systems use sun-seeking sensors or a computer to calculate the position of the sun and move the array accordingly using motors and gears. There are two main types: single-axis trackers and dual-axis trackers. A single-axis tracking system moves the solar modules through one axis from east to west to follow the sun's path over the course of the day, whereas a dual-axis tracking system follows the sun's path along two axes: like the single-axis tracking system it will follow the sun's path throughout the day, and it also makes adjustments in tilt angle to account for changes in the sun's altitude over the year (i.e. in winter the sun is much lower in the sky so the modules will be tilted at a steeper angle).

The disadvantage of sun-tracking systems is that they have many moving parts, which increases maintenance costs over the life of the system. There is also a higher risk of mechanical failure than in fixed tilt systems because of these additional moving parts. The calculations to show the increased power output from any PV array using a single- or dual-axis tracker should be done so that the financial gain from the tracking system can be offset against any additional material, installation and maintenance costs. Tracking systems can also utilize concentrators to maximize the benefit of tracking. These systems are highly irregular outside utility and academic installations.

Figure 6.16 Sun-tracking system: the arrays rotate throughout the day to face the sun

Figure 6.17 The tracking hardware can be seen on the back of this array

Source: Frank Jackson

Wind loading

Wind loading describes the wind forces experienced by a solar module, including a suction or uplift force on the module when the wind blows across the array and downward or lateral stresses on the module caused by strong winds. A PV array is not mounted safely unless it can withstand the wind loading forces expected at the site. Wind speeds vary widely throughout the world and so it is important to ensure that the module and mounting system selected are suitable for use at the site. The module's installation manual or data sheet specifies the module's maximum load rating and should be consulted before final module selection.

Warranty Information   10 years
Standards and Compliance   NZS 3604-1999
AS/NZS 1170.2-2002

Contact Information

+64 9 889 0500 · NZ   Email: info@powersmartsolar.com.au
+61 2 8011 3015 · NSW   Web: www.powersmartsolar.com.au
+61 3 9010 5003 VIC   Skype: powersmartsolar
+61 7 3103 0156

Figure 6.18 Example of a mounting system information sheet, displaying compliance to Australian and New Zealand wind loading standards

Source: Powersmart Solar

The effect of wind loading will also greatly depend on the height and tilt angle of the PV array and the exposure of the module. In order to minimize wind loads, solar panels should be installed parallel to the prevailing wind direction and if roof-mounted, away from corners and roof edges. The array structure must be designed and installed to meet the site's wind speed requirements so that the array will never be lifted or blown off the roof. This will be highly dependent on local standards (see resources section), which reflect local design conditions.

In general, an array frame or structure is designed by the manufacturer's or supplier's structural engineer to meet the specific wind loadings for the complete array's installation in the conditions of the country/local area where it is sold. The design should ensure that the metal used in the frame is sufficiently strong to withstand the wind loading forces, so calculating the wind loading of the structure is usually not a problem. However the critical factor is the point of attachment of the mounting structure to the roof. A critical design consideration is to ascertain the strength of the point of attachment and from there, calculate how many attachments are required to securely fix the mounting structure to the roof. Manufacturers generally sell mounting systems as a kit containing the rails and a quantity of fixing brackets required for the region where the array is to be installed. For example; in tropical areas more roof fixings and mounting brackets are required due to possible hurricane, cyclone or typhoon conditions. A mounting structure should be compliant with local wind exposure codes and it may be necessary to use the services of a qualified engineer and obtain a structural certificate to ensure that it meets the relevant requirements.

Lightning protection

Module frames and mounting structures are almost always metallic and therefore good conductors of electricity. The mounting structure generally needs to be grounded (or earthed), although it is important to check the relevant national electrical codes and standards as such regulations differ widely from country to country. Incorrect lightning protection may increase the risk of injury or damage to the PV array or building in the event of a lightning strike. Lightning protection is discussed in more detail in Chapter 5.

7

Site Assessment

Conducting a site assessment or site survey is an important step in the design and installation of a system. During the site assessment the installer should collect all the necessary information required to optimize system design and plan for a time-efficient and safe installation. A site assessment aims to determine the location of the PV array, the roof specifications, the amount of shading, the available area and other considerations.

Location of the PV array

In most urban areas the array is located on the roof of a building, or in cases where there is a large, clear area of ground that will not be shaded (by trees or nearby buildings) it may be desirable to use a ground-mounted system. Mounting structures are discussed in Chapter 6. There are many options available and these often depend on the angle and orientation of the roof or ground.

Roof specifications

- Orientation: as discussed in Chapter 2, the ideal orientation is where a module receives maximum sunlight (this is true south for the northern hemisphere or true north for the southern hemisphere). Unfortunately when a PV array is installed on a roof its orientation is governed by the direction of the roof. Using a compass and magnetic declination data (as discussed in Chapter 2) installers should determine the orientation that the roof is facing and its bearing from the ideal orientation. The orientation of the roof will be the same as the orientation of the modules and will be required for energy yield calculations.
- Tilt angle: in most systems the tilt of the modules will follow the tilt angle (or pitch) of the roof. The tilt angle should be measured using an inclinometer or an angle finder; it may also be available on architectural drawings of the building. The optimum tilt for a system is equal to the latitude of that location. In cases where the pitch of the roof is not equal to the optimum tilt angle the PV array's energy yield will be affected.

Once the tilt and orientation angles of the PV array have been determined, the designer may want to calculate their effect on energy yield. This is normally done using data tables rather than by hand; in the US, data tables are available

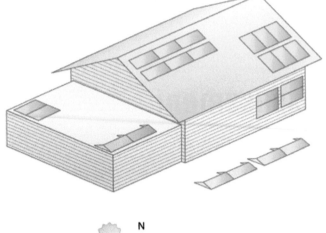

Figure 7.1 Modules may be laid flat on the surface or tilted on rack mounts, which are particularly useful for flat roofs

Source: Global Sustainable Energy Solutions

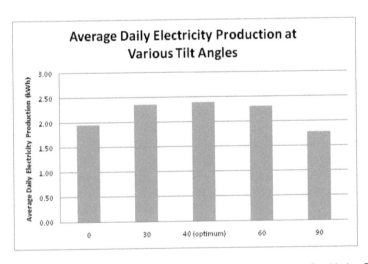

Figure 7.2 This image shows the orientation of a module installed in the southern hemisphere; in the northern hemisphere the optimal orientation is due north

Source: Global Sustainable Energy Solutions

Figure 7.3 Average daily electricity production for a PV system installed in Stockholm, Sweden. This graph shows the effect of tilt angle on energy yield. As can be seen the average daily energy yield at the optimum angle is higher. Over the whole year this makes a significant difference to the output

Source: PVGIS © European Communities, 2001–2008

from the National Renewable Energy Laboratory (NREL), and the PVGIS online tool can be used to calculate the effect of tilt and orientation on power output for any location in Europe and Africa (see Chapter 15).

Is the site shade-free?

As explained in Chapter 3 shading on a PV array can significantly decrease its output. Some sources of shading such as dust, dirt and bird droppings are unavoidable and must be cleaned off regularly. Any permanent source of shading needs to be identified during the site survey. Potential sources of shading may include:

- Trees and vegetation: it is important to bear in mind that trees that do not shade the PV array at the time of the site assessment will grow and may shade the PV array in a few years, so this should be discussed with the owner before installation. The owner may agree to prune the tree regularly and ensure it does not shade the array. If this is not possible, for example, if the tree is on a neighbouring property, another location should be considered, or the neighbour may be consulted.
- Other buildings, including neighbouring properties or buildings on site. Be aware that new buildings may be constructed, shading areas currently suitable for arrays.
- Components of the building on which the array is to be installed; chimneys, TV aerials, satellite dishes and other sections of roof could all shade the

Figure 7.4 Poorly chosen site where trees are shading the PV array

Source: Global Sustainable Energy Solutions

Figure 7.5 In this installation the chimney is shading the array

Source: SMA Solar Technology AG (not an SMA installation)

array if it is not located properly. The user should be made aware that future rooftop installations should not shade the array at any time.

- The natural landscape: mountains or hills may shade a solar array, particularly when the sun is low in the sky.

The PV array should be installed where it will not be shaded during the day. Shading will decrease power output and may damage the modules over their lifetime. Some shading may be acceptable very early or late in the day; however, this decision should be left to a qualified PV system designer.

It is also important to consider local regulations regarding shading or 'solar access'. Many US states have enacted laws to prevent people from growing trees or extending their houses in ways that would shade an existing PV system on a neighbouring property. Many locations have no such law, but good communication between neighbours can help to ensure that they do not make changes to their property that will adversely affect the PV system.

Aside from observing the physical objects on site it is also important to conduct a thorough shading analysis. The sun does not stay at the same point in the sky throughout the year so even though an object may not be shading the proposed location on the day of the site assessment does not mean it won't be shading the location at some other time.

There are a number of tools available to estimate shading on a site throughout the year. The following examples only require one visit to estimate the solar resource for the entire year:

Figure 7.6 Solar Pathfinder in use

Source: Global Sustainable Energy Solutions

Solar Pathfinder

A Solar Pathfinder is used to determine the extent of shading at the intended PV installation site. It is used to identify objects which will cause shading and the times of year when shading will occur.

A Solar Pathfinder requires a sunpath diagram for the latitude of the location. This diagram is then placed inside the Pathfinder, the diagram is rotated so that the white dot on the rim lines up with the magnetic declination given for that location and then the entire Pathfinder must be rotated so that the needle at the base points to magnetic north. The transparent dome is then placed over the top of the Pathfinder.

When the Pathfinder is placed at a given location, the objects surrounding it will cast a shadow on the dome (and the diagram inside) and the outline of these shadows is then traced onto the sunpath diagram in wax pencil to determine when the array in that location will be shaded. If no shadow is cast on the Solar Pathfinder the site will be shade-free all year.

The horizontal arcs represent the sun's path across the sky for each month of the year. In Figure 7.8 the large tree to the north of the site will shade the array at some point of the day in every month except for November, December and January. Vertical lines represent the hours of the day: throughout the year this array would be shaded before 8.00am. Demonstration videos showing how to set up and use a Solar Pathfinder are available on www.solarpathfinder.com/video.

Though the Solar Pathfinder is an excellent tool, it cannot always be used effectively. If the building on which the solar array will be located has not yet

Figure 7.7 Solar Pathfinder, showing trees to the north and to the right shading the site

Source: Global Sustainable Energy Solutions

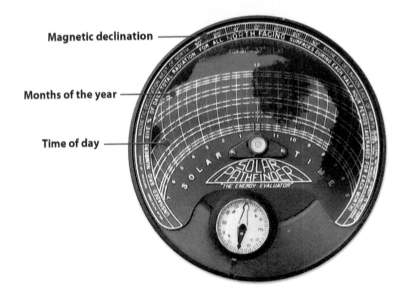

Figure 7.8 Example of a sunpath diagram with the shading traced on it; it is important to use the correct sunpath diagram for the latitude of the site

Source: Global Sustainable Energy Solutions

been constructed, then the Solar Pathfinder cannot physically be located at the same height as the proposed array. A full assessment of the effects of shadows on the proposed array cannot be undertaken until after the construction. Solar Pathfinder software is also available: rather than tracing the shadows onto the sunpath diagram, a photo can be taken of the shaded dome and uploaded to the computer. This software is also able to adjust the shading analysis at ground level for a given height – allowing the Solar Pathfinder to be used in circumstances where it cannot be physically located at the same height as the proposed array.

Solmetric Suneye

The Solmetric Suneye is a digital tool which is positioned flat on the intended installation site. The user must specify the latitude of the location. It takes a picture using a fish-eye camera lens of the horizon and analyses the effect of surrounding objects. Accessories, such as GPS units, can be purchased to improve the accuracy of this analysis. The Suneye also has a level and inclinometer function to measure roof angle and orientation. More information is available at www.solmetric.com/sosu.html.

HORIcatcher

The HORIcatcher is another widely used digital tool. A digital camera is mounted above a horizon mirror; the photos taken are then uploaded and used with associated software to calculate the irradiation available at the location. More information can be found at www.meteotest.ch.

iPhone apps

There are also a number of iPhone applications that can be uploaded from Apple Inc. and used to trace the horizon and provide site analysis in a number of different ways:

- The Solar Checker accesses a database to determine the solar radiation for a given GPS location. The magnetic compass in the phone is then used to measure the roof's orientation and its inbuilt tilt sensor to determine the angle. From these values the Solar Checker uses a mathematical model to estimate the annual energy yield.

Figure 7.9 Solmetric Suneye

Source: Solmetric Corporation

Figure 7.10 HORIcatcher in use

Source: Meteotest, Switzerland (www.meteotest.ch)

Figure 7.11 Screenshots of the iPV app

Source: Solmetric Corporation

- Solmetric iPV Solar App is owned by the same company that manufactures the Solmetric Suneye. While inexpensive, it is not as powerful as the Suneye tool. The iPV app is able to measure the orientation and angle of the roof, estimate the solar radiation available and generate a site survey report. Further information is available at www.solmetric.com/solmetricipv.html.

Software packages

As well as software available with individual tools, there are other simulation programs to analyse the effect of shadowing on the proposed array. To use these effectively, the physical dimensions of the object in relation to the location of the proposed array must be determined. A 3D computer model of the site and proposed array is built and then the program is able to simulate the shadows cast by those objects. Examples of this type of software include:

- Shadow Analyzer: more information available at www.drbaumresearch. com/prod38.htm.
- Ecotect: this architectural tool is also able to conduct thermal and energy analysis of buildings; more information available at: http://usa.autodesk. com/adsk/servlet/pc/index?id=12602821&siteID=123112.
- PV*SOL: more information available at www.valentin.de/produkte.
- PVsyst: more information available at www.pvsyst.com.

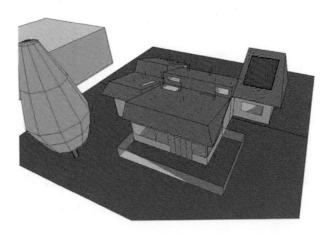

Figure 7.12 Screenshot from architectural software showing a 3D model of a house with PV array

Source: Global Sustainable Energy Solutions

Available area

The first design constraint on the system is how many modules can physically fit into the shade-free location identified. In some cases the roof will obviously be larger than the area required for the PV array. However, when a roof has many smaller sections or protruding objects such as chimneys it may be necessary to calculate the available area suitable for the PV installation. The available area is very important in designing the array and choosing the module, as discussed in Chapter 10.

To determine the available area, the following steps are taken:

1 Measure the available roof space; measurements can be obtained using a tape measure or from existing architectural drawings of the building. In this book the distance between the gutter (or bottom edge) of the roof and the ridge is defined as the 'width' and the horizontal distance along the roof as the 'length'.
2 Determine the 'edge zone'; depending on the mounting frames chosen, a margin or 'edge zone' should be left around the perimeter to make sure that the installation and any further maintenance can be safely carried out. The size of the 'edge zone' is usually covered by local codes, i.e. in Australian the 'edge zone' is recommended to be 20 per cent of the available area.
3 Determine the maximum number of PV modules that can fit in the available area; the actual number of modules that can fit on a roof is limited by the length and width of the module compared with the length and width of the available roof space. A module is generally rectangular in shape and has two fixed dimensions, length and width. Modules can be installed either mounted in landscape (length parallel to the length of the roof) or in portrait (length parallel to the width if the roof), as shown in Figures 7.13 and 7.14.
4 If space is an issue, the orientation allowing the installation of the most modules must be calculated. If there are non-rectangular areas, they should be broken down into rectangles, and then the calculations use the following methods, multiple times.

Figure 7.13 Landscape orientation

Source: Global Sustainable Energy Solutions

Figure 7.14 Portrait orientation

Source: Global Sustainable Energy Solutions

Portrait installation

Traditionally PV modules were mounted in portrait, with mounting rails installed across the roof. First, the number of rows that can be installed must be calculated by dividing the width of the roof by the length of the module. The edge zone must be factored in and deducted from the width of the roof before the calculation takes place.

The number of columns can then be calculated by dividing the length of the roof minus the edge zones by the width of the module.

The number of rows and columns must then be multiplied to obtain the maximum number of modules for that roof area.

Landscape installation

Many mounting systems can now be used to mount modules in landscape. First, the number of rows that can be installed must be calculated by dividing the width of the roof by the width of the module. Remember to deduct the edge zone from the width of the roof.

The roof length should be reduced by the edge zone size and then the number of columns is calculated by dividing the length of the roof by the length of the module. The number of modules along the length and the width are then multiplied to determine the total.

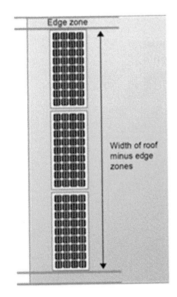

Figure 7.15 Example of calculating the number of rows for a portrait installation

Source: Global Sustainable Energy Solutions

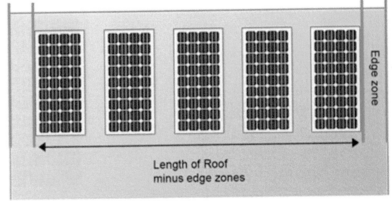

Figure 7.16 Example of calculating the number of columns in a portrait installation

Source: Global Sustainable Energy Solutions

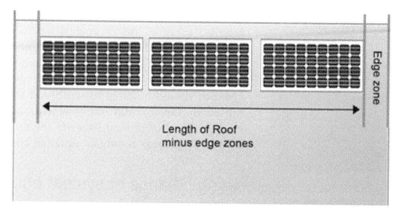

Figure 7.18 Length of roof minus edge zones

Source: Global Sustainable Energy Solutions

Figure 7.17 Width of roof minus edge zones

Source: Global Sustainable Energy Solutions

5. While the total number of PV modules that can be mounted on an available roof space is important to consider, there are other constraints. The final number will depend on the choice of inverter and the specific number of modules that can be connected in series and parallel to that inverter, as discussed in Chapter 9.

Energy efficiency initiatives

The designer should provide the system owner with a list of recommendations to reduce their energy usage so that the proposed grid-connected PV system can meet all or a larger portion of their daily electrical energy requirements.

Health, safety and environment (HSE) risks

There are several key risks associated with PV installations; these include working at heights, manual handling (lifting and moving heavy objects) and risks from working with electricity. Installers should conduct a risk assessment of each site and determine the site-specific risks in their site assessment. They should review this work before commencing the installation.

Local environment

Any features of the local environment that will affect equipment selection must be identified; these may include:

• Local temperature range: equipment will only operate within a specified temperate range, which is given on equipment data sheets. It is also important to know the local temperature range when designing the PV array (as discussed in Chapter 9).

Figure 7.19 Badly located inverter; the system owners will have great difficulty accessing information from this inverter

Source: Global Sustainable Energy Solutions

- Corrosive conditions such as salt in the atmosphere: when a system is to be installed in an environment exposed to salt, such as within 1km of the coast, suitable salt-tested modules should be chosen that are certified to IEC 61701 – Salt mist corrosion testing of photovoltaic (PV) modules.
- Snow: in regions subject to heavy snow loads, the solar module to be installed should be rated to withstand the increased downwards pressure caused by the potential accumulated snow loads on the face of the PV module. The higher load capacity rating of 5400Pa for PV modules to be used in these areas is only an optional compliance under IEC 61215.

Locating balance of system equipment

The location of the solar array has already been covered in this chapter. While on site, the location of all other equipment must be determined. This includes the location of:

- PV combiner box (if required);
- Inverter: the inverter should be located somewhere where it is easily accessible, protected from direct sunlight and well ventilated.

The location of all equipment should be marked on an architectural/building drawing if available; if this is not available the person undertaking the site visit should prepare a site sketch. Digital photos of the site and the proposed location of each piece of equipment should be taken as a record.

Figure 7.20 Well-located BoS components are protected, well-ventilated and neatly installed

Source: Global Sustainable Energy Solutions

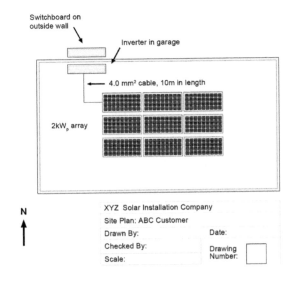

Figure 7.21 Example of site plan for a PV installation

Source: Global Sustainable Energy Solutions

Site plan

Before installing the system, installers need to complete a site plan detailing any buildings, solar obstructions, access pathways and the physical layout of the site. It is necessary to overlay the designed system onto this drawing. The site plan should include dimensions such as roof lengths, cable lengths, position of components and distances between buildings.

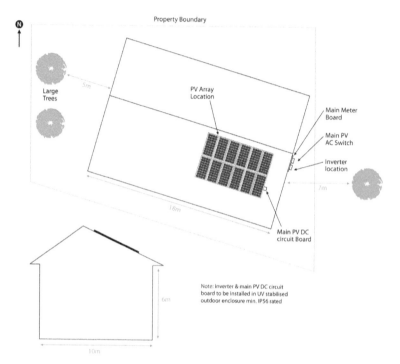

Figure 7.22 Example of a site plan

Source: Global Sustainable Energy Solutions

Box 7.1 Grid-connect site assessment form

Type of installation:

New system Upgrade
(details)...

Array mounting specifications:

Type of mounting:
Roof Ground Pole Other
(details)...

Roof specifications:

Orientation:................ (see section on orientation in this chapter)
Angle:.................. (see section on angle in this chapter)
Approx area (m²):.........
Is there shading of roof where array is to be installed?   N   Y
If there is shading, when?

Month	Shading start (hh:mm)	Shading finish (hh:mm)	% Array area shaded (max)
Jan			
Feb			
Mar			
Apr			
May			
Jun			
Jul			
Aug			
Sep			
Oct			
Nov			
Dec			

Grid connection specifications:

Existing service panel / switchboard (SB)
     SB: Adequate   Upgrade required
     Location details:..
     Sub-panel (if applicable) OK Upgrade required Dist. from SB: .... m
     Inverter mounting:
     Timber Brick Steel Other:...........................
     Existing meter type: Utility:
     Number of modules:........................
     Layout (e.g. 4 × 2 Portrait) :...
     Proposed array size (kW):...................................

Additional site information:

....
....
....
....

Note any additional information about the site that may affect the installation, e.g.:

- Environmental factors, i.e. snow, high winds, salt;
- Safety issues, i.e. uneven ground, access to roof, limited working space;
- Any changes required, i.e. pruning trees.

8
Designing Grid-connected PV Systems

Following a successful site assessment at which all necessary information is collected, the system can be designed. The selection and design of system components are covered in this chapter. Most of the equipment described here provides safety as well as guaranteeing system longevity. It is important to note that in most countries all major points of system design are covered by national codes and standards, and designers need to familiarize themselves thoroughly with these.

Design brief

Before any components are selected it is necessary to consult the client regarding their needs and expectations from the system. The amount of money the client is willing to pay for the system will affect the size of the system, as will the metering system. Different metering systems are discussed in Chapter 5: when gross metering is used, the system installed is often as large as possible to capitalize on the gross metering returns; when net metering is in place, the consumer may choose a smaller system and decide to invest more money in energy-efficiency measures on site so as to minimize local electricity consumption and maximize the energy exported.

Existing system evaluation

When designing a system, it can be useful to approach PV system owners and installers in the area to see what location-specific issues they have encountered, if any, and how they have dealt with them. Some regions experience severe weather, others may have destructive wildlife and still others may have high humidity throughout the year. These phenomena may affect the longevity of your system if they are not considered. Speaking with as many experienced local technicians as possible helps designers set up a system that will perform well under local conditions as well as complying with local codes and standards.

Choosing system components

An important part of system design is choosing components that are appropriate for the system and environment. This chapter lists important factors to consider when choosing system components, while Chapter 9 discusses matching the array and inverter. Chapter 6 covers choice of mounting system.

Modules

Choosing and ordering the appropriate modules is very important. They are usually the most expensive part of the system and it can be a costly error if the incorrect product is chosen. This choice should not only be governed by the performance, efficiency and cost of the module but also by the conditions under which it will operate. The factors affecting the choice of module are discussed below, while sizing the system is outlined in Chapter 9.

Particular environmental conditions that will affect module choice should have been identified during the site assessment and are briefly summarized below:

- Local temperature range: Installations in a hot climate require modules with low temperature coefficients in order to minimize the decrease in power output caused by an increase in temperature. PV modules' operating temperature range is specified on the data sheet – it is important to ensure that the installation location's temperature range is within that of the PV module selected.
- Atmospheric salt in coastal environments: Modules installed within 1km of the coast should comply with IEC 61701 – Salt mist corrosion testing of photovoltaic (PV) modules.
- Heavy snow: For snowy areas it is important to install a module that has a load capacity rating of 5400Pa; this will be specified on the data sheet.

The aesthetics of the module may also be important to the system owner. A system with identical modules will generally look better. There often aren't many choices regarding colour, but new modules are being developed with aesthetics in mind.

Finally, the national codes and standards may outline requirements for PV modules, e.g. a module should not be sold in the European Union unless it has the mandatory CE performance mark.

Mounting structure

Once the system's modules have been selected, designers must determine how they will be mounted. This decision depends not only on the type of module but also on the installation site and the local environment. This information is covered in Chapter 6. Environmental characteristics such as frequency of heavy rainfall, proximity to the coast (and hence corrosive environment) and local wind-loading requirements play an important role in deciding what type of mounting system to use. The designer may also have to consider the customer's aesthetic requirements for the PV system.

Figure 8.1 There are several module types on the market that are all (or mostly) black. It should be kept in mind that crystalline modules with a black backing sheet will have a higher effective cell temperature and therefore lower power output than the same module with a white backing sheet

Source: SunPower

Inverters

The main types of inverters, as discussed in Chapter 5, are: central, multi-string, string and modular. When selecting an inverter it is important to consider the following:

- The peak rating of the PV array: the maximum PV array rated power an inverter can handle will be given on the data sheet. As a rough guide, this should be similar to the peak rating of the PV array, but more complex calculations are required to ensure safety and these are shown in Chapter 9.
- Whether all the PV modules have the same tilt angle and direction: if the modules are to be at different tilt angles or facing different directions, it is generally better to divide the system into a number of different strings and use a multi-string inverter or a single string inverter on each string. Typically

it is more expensive to use several smaller inverters instead of one large inverter, although the higher energy yield advantage may outweigh the additional capital costs.

- The efficiency of the inverter: most modern inverters have similar efficiencies, but it may sway the system designer's choice. Transformer-less inverters are generally more efficient and are now widely used in Europe and Australia. However, they are not as common in the US as they were not permitted by the National Electric Code until 2005.
- Inverter location: the inverter's data sheet specifies an electronics protection rating. Inverters that are to be installed outdoors should be impervious to water and dust, an electronics protection rating of Ingress Protection 65 (IP65) or National Electrical Manufacturers Association 3R (NEMA3R) is evidence of this. An explanation of the different NEMA rating can be found at www.nema.org and the IP ratings are outlined in IEC 60529 Degrees of protection provided by enclosures (IP Code) available at www.iec.ch. The IP rating standard is internationally recognized and commonly used in Europe and Australia while the NEMA standard is the most commonly used in the US.
- The capital costs of the different inverters.
- The average annual expected energy yield.

The designer needs to determine the advantages and disadvantages of each solution with respect to capital costs and system performance. These considerations and the associated costs should be explained to the client and the final decision made in consultation with them.

There is a variety of software available to assist in the design of PV systems. Most of this software stores and accesses information on real market products including PV modules and inverters, giving users hundreds of different options regarding manufacturer and model. Designers can use software to simulate different systems using different products and compare their outputs in order to choose the system which best fits their design brief. The software can then be used to size the array to the inverter, saving designer's time which would otherwise be spent doing complex calculations (see Chapter 9). Most software is able to access meteorological data and can therefore simulate output by factoring in the system's location. Alternatively users may be able to enter information regarding local solar resource and shading. Examples of commonly used software are PV*SOL, PVsyst, SMA's Sunny Design and Homer, all readily available on the web (see Chapter 15). Note that of the programs listed, Sunny Design is the only one that is free.

Cabling

Cabling is covered in further detail in Chapter 10. It is important that the cable chosen is appropriate for use in a PV system (i.e. UV-resistant for outdoor use) and appropriately sized for the expected currents and voltages of the system. Cable sizing is covered extensively by local codes, which should always be followed as there is a lot of variation between different countries. The sizing of the cable will often vary with its position (i.e. a string cable may be sized

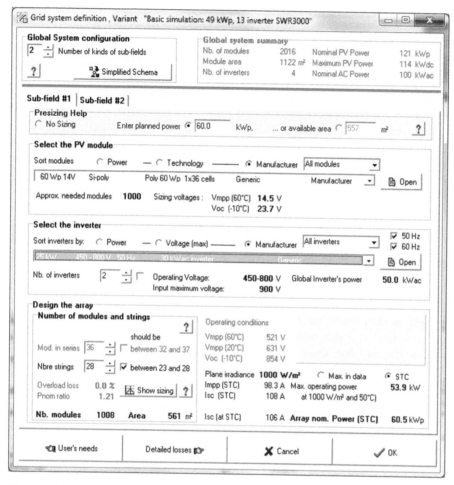

Figure 8.2 Screenshot from PVsyst

Source: PVsyst

differently to an array cable). The cables in an installation must be sized correctly to ensure that there is not an excessive voltage drop (power loss) between the array and the inverter and that the cables are sufficiently large to conduct the current (see Chapter 10). If the current passing through the cables is greater than the cables' current-carrying capacity, the cable will overheat, which can damage the circuit components. Cables should be kept as short as possible and must also be rated for the temperature and environment (i.e. UV-resistant) within which they will be installed.

Voltage sizing

Each cable will have a maximum voltage rating (available from the manufacturer) and this must never be exceeded; sizing standards vary between countries.

In Australia, DC cables should be rated for at least $1.2 \times V_{oc}$ of the component (i.e. the string, sub-array or array).

Current sizing

Cables will also have a maximum current rating which may not safely be exceeded. For example, in the UK and Australian standards, DC cables in a PV system must be rated to carry at least $1.25 \times I_{sc}$. This means that string cables must be rated to $1.25 \times I_{sc}$ of the module and the DC main array cable must be rated to carry $1.25 \times I_{sc}$ of the array (I_{sc} of the array = I_{sc} of the module × number of parallel strings in the array). In the US the standard is slightly different and DC cables must have a minimum current rating of $1.5625 \times I_{sc}$, i.e. the string cables will be rated at $1.5625 \times I_{sc}$ of the module and the DC main array cable will be rated at $1.5625 \times I_{sc}$ of the array.

Monitoring

The choice of inverter might also include the possible monitoring options, such as a wireless display that the customer can use inside the home to monitor system performance, or the ability to monitor electricity production and consumption at a site using multiple inverters at a large facility. Bluetooth is also a feature that some customers are now requesting, allowing wireless monitoring on a computer, PDA or mobile phone. Monitoring is covered in further detail in Chapter 12.

System protection

The design of the protection system is important to ensure the safe operation of the PV system. There are many potential reasons for a PV system to fail. Natural threats such as lightning, flooding or strong winds can destroy system components or cause dangerous system operation. National codes typically require many types of protection in PV systems for safety; over-current protection, lightning and surge protection and a means of disconnection are all common requirements. The details of what protection measures are required are set out in national codes which should always be followed. The hardware commonly used as system protection devices is covered in Chapter 5.

Over-current protection

All circuit over-current protection, both AC and DC, is designed to protect the components and cables in which it is installed from damage due to overload currents or short circuits. The size of the over-current protection is determined by the type of device used and the maximum current which can be safely passed through the circuit elements. National codes will outline how over-current protection should be designed and sized. In the UK and Australia (and many other parts of the world) all sizing is based on the short-circuit current of the array. The following points are key reasons why the short-circuit current of the array is used:

- A PV module is a current-limited device, i.e. the highest current it can produce is its I_{sc}.
- The short-circuit current is the maximum current a PV module can produce at a given temperature and irradiance.
- The maximum current that an array can produce is the sum of the short-circuit currents of each string in the array.

Since the short-circuit current can vary depending on the temperature of the solar modules and the irradiance, these two important factors should be taken into account when designing photovoltaic systems.

Fault-current protection

It is important to note that there is a limit to the amount of current that can pass in either direction through a PV module without damaging it. This is known as the 'maximum series fuse rating', 'reverse current rating' or 'over-current protection rating' and is stated on the module data sheet. If a fault occurs within one string it is possible that the current from the other strings will feed into the faulty string. If this current exceeds the maximum series fuse rating the module is likely to suffer damage. This is normally only a problem for arrays with multiple strings, i.e. the sum of the currents of all the non-faulty strings exceeds the maximum series fuse rating. To prevent this occurring, string fuses or miniature circuit breakers (MCBs) are often used, and these must be rated for use in a DC PV system.

To determine whether fault-current protection is necessary, designers need to know the short-circuit current of the array and the modules' maximum series fuse rating. National codes normally specify what is required and should always be consulted. For example, in the UK string fuses are required whenever the reverse current rating of the modules (I_r) is less than the combined short-circuit current of all the strings minus one multiplied by 1.25, shown as follows: $I_r < \{I_{sc} \times (\text{number of strings } -1) \times 1.25\}$. As a rule of thumb, string fuses are required in all arrays of four or more strings, and the fuses used must meet the following requirements:

- Fuses must be fitted to both the positive and negative string cable.
- Fuses must be rated for DC (see Chapter 5 regarding the difference between AC and DC fuses).
- Fuses must be rated: ($V_{oc} \times$ number of modules in the string $\times 1.15$).
- Fuses must have a tripping current (I_{trip}) of less than $2 \times I_{sc}$ or less than $2 \times$ current-carrying capacity of the string cable (whichever value is lowest should be used).
- If the system does not have string fuses, then the string cable should have a minimum current rating of $I_{sc} \times$ (number of strings -1) $\times 1.25$.

National codes and regulations must be consulted.

> ## Box 8.1 Fault-current protection example
>
> National Australian Standards are slightly different to those in the UK. In Australia fault-current protection is required when the combined short-circuit current of all strings in the array minus one is greater than the reverse current rating of the modules, i.e. $I_r < I_{sc} \times$ (number of strings −1).
>
> Example: Is fault-current protection required for an array with 5 strings, each module has a short-circuit current of 5.69A and a maximum series fuse rating of 15A?
>
> First calculate $I_{sc} \times$ (number of strings −1) = 5.69 × (5 − 1) = 22.76A
>
> 22.76A is larger than the maximum series fuse rating of 15A, so fault-current protection is required.
>
> When the potential fault current of the array is less than the modules' reverse current rating, then over-current protection is not required, but the cables used throughout the array should be appropriately sized to handle the maximum possible fault current.

Lightning and surge protection

Lightning protection and surge protection requirements are usually included in national codes and regulations, and can even differ from region to region within a country. For example, in Australian/New Zealand Standard AS/NZS 1768: 2007, surge arresters are required when the array feeds supply to critical loads (e.g. telecommunication repeater stations) or if the PV array has a rated capacity greater than 500 watts. A surge arrester is commonly found in commercial PV inverters. In systems where this is not the case, protection is recommended through the use of metal oxide varistors (MOVs). According to AS1768, an MOV should be selected with a continuous operating voltage of greater than $1.3 \times V_{oc}$ of the array and a maximum discharge current greater than 5kA. If a lightning protection system is in place the PV system may need to be integrated into it – how this is done should be confirmed as consistent with prevailing national codes and regulations.

Grounding/earthing

Grounding/earthing systems are specified by local codes and standards. The array mounting structure and the array itself are considered separately. Grounding/earthing the mounting structure is generally done for lightning protection and to provide a path for fault currents to flow. Whether an array conductor is required to be grounded/earthed and in what manner is a more complex matter. Following the recommendations of local standards and codes is necessary.

String fuse protection: When is it required and when not?

Example:

Module:	**Sunshine Model S185**
Isc:	**5.5A**
Maximum series fuse rating*:	**15A**

Case One: 2 Parallel Modules : One is shaded or faulty

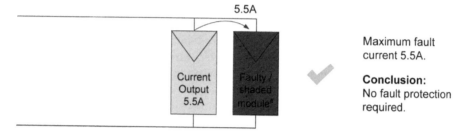

Maximum fault current 5.5A.

Conclusion:
No fault protection required.

Case Two: 3 Parallel Modules : One is shaded or faulty

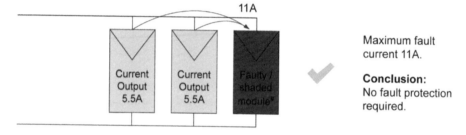

Maximum fault current 11A.

Conclusion:
No fault protection required.

Case Three: 4 Parallel Modules : One is shaded or faulty

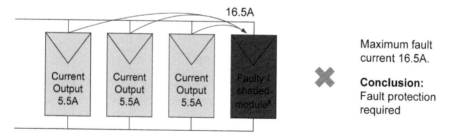

Maximum fault current 16.5A.

Conclusion:
Fault protection required

Therefore: array must be protected in each string because of the unknown position of fault.

* Module Reverse Current Rating is also known as Maximum Series Fuse Rating
# Assume 0A current output

Figure 8.3 Example of when string protection is/isn't required by Australian Codes
Source: Global Sustainable Energy Solutions

Mechanical protection

As PV arrays are often located as high as possible to prevent shading and interference from external agents, they are often subject to high levels of wind loading. As such, the support structure for these arrays should conform to the local building code and any national standards or regulations regarding wind loading. Wind loading is covered in Chapter 6.

Array protection

Array protection is dependent on local codes and varies by country. As there are no other sources of DC current (i.e. no batteries are connected), array fault-current protection is not typically required in grid-connected PV systems without batteries. National codes do, however, normally require the installation of a load-breaking disconnection device between the array and the inverter. This is called the PV array main disconnect/isolator. It is normally located near the inverter and is typically required to be lockable so when maintenance is being carried out the array can be safely disabled. This disconnect/isolator must be rated for DC; a disconnect/isolator rated only for equivalent AC use is unsafe. Safe voltage and current parameters, as well as whether the switch must disconnect one DC line or both, are specified in the relevant code; for example, Australian/New Zealand Standard AS/NZS 5033: 2005 requires a lockable switch that disconnects both lines (double-pole) and has a rating of $1.2 \times V_{oc}$ to be installed.

Sub-array protection

Arrays may be divided into sub-arrays for several reasons, e.g. if two parts of the array are installed in separate areas. If this is the case, sub-array protection may be required by local codes, which will specify the rated trip current required for over-current protection for the sub-array, normally related to the sub-array's short-circuit current (I_{sc}). Local codes may also require the sub-array to have a load-breaking disconnection device to isolate it from the rest of the system.

Extra low voltage (ELV) segmentation

Another reason for dividing arrays into sub-arrays may be so that the module string can be broken down into ELV segmentation. Most PV arrays operate with their voltage in the range 120–500V DC. It may be desirable to split the array into ELV sub-arrays with V_{oc} of less than 120V in order to make the array safer and reduce the risk of electrocution. National codes generally address ELV segmentation; for instance in Australian/New Zealand Standard 5033: 2005, ELV arrays do not require fault-current protection or disconnection devices on individual strings, whereas LV arrays must have a suitable method of disconnection into ELV segments.

9

Sizing a PV System

In Chapter 8 the choice of system components is discussed. While the client's requests and expectations of their system will give the designer a rough idea of the system size, it is important to carry out detailed sizing calculations to ensure that the array is matched to the inverter's input specifications and that all system components are appropriately sized to suit the site-specific conditions.

In general, an inverter should be chosen so that the maximum power output from the designed PV array (given in Wp) matches the inverter's maximum array power input (given in Wp). Selecting an inverter with a power input that is larger than the array's output is known as over-sizing the inverter and should be avoided as it will reduce the inverter's operating efficiency and hence reduce overall power output. The inverter's actual operating efficiency should be as high as possible and this is described as that point where the power of the PV array is matched as closely as possible to the inverter's rating in watts. The closer these two factors are to each other, the higher efficiency will be. It is not enough that the inverter's operating efficiency is as high as possible; the designer needs to ensure that the inverter and PV array match in terms of voltage, current and power to ensure a safe and efficient PV system. System sizing is often done using computer programs such as PV*SOL or PVsyst, but the basic calculations need to be understood. This chapter outlines system sizing using two fictitious examples: one in Berlin, Germany, and the other in Sydney, Australia. This chapter discusses system losses and their primary sources and concludes with examples of system yield calculations factoring in array size, local solar resource and system losses. The sizing and de-rating methodology in this chapter is widely used, but it is not the only method available and local standards and codes should always be consulted as they may specify how these calculations should be undertaken in order for the system to be compliant in that location.

Matching voltage specifications

There are two voltage specifications that need to be met. The first is that of the module itself: most manufacturers specify a maximum system voltage on their data sheet. A PV array's voltage must not exceed the maximum system voltage for the modules it is using.

The second voltage figure is the maximum input voltage of the inverter which must not be exceeded: this figure is normally lower than the module's maximum system voltage and hence of greater concern. This figure is highlighted in the following extract from an inverter data sheet (Figure 9.3).

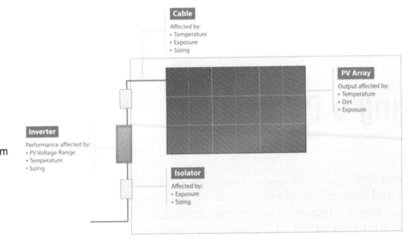

Figure 9.1 Basic PV system components showing the major components and factors which affect their performance

Source: Global Sustainable Energy Solutions

Electrical Data		
Measured at Standard Test Conditions (STC): irradiance of 1000/m², air mass 1.5g, and cell temperature 25°C		
Peak Power(+/- 3%)	P$_{MAX}$	210W
Rated Voltage	V$_{MP}$	40V
Rated Current	I$_{MP}$	5.25A
Open Circuit Voltage	V$_{OC}$	47.7V
Short Circuit Current	I$_{SC}$	5.75A
Maximum System Voltage	IEC	1000V
Temperature Coefficients		
	Power	-0.38%/°C
	Voltage (V$_{OC}$)	-136.8mV/°C
	Current (I$_{SC}$)	3.5 mA/°C
Series Fuse Rating		15A

Figure 9.2 Extract from a 'sample' module data sheet highlighting the 'maximum system voltage' figure allowed for use with that solar module

Source: Global Sustainable Energy Solutions

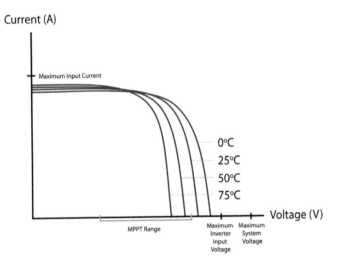

Figure 9.3 The current and voltage of a PV cell or module varies with temperature, so it is important that the PV system operates safely at all temperatures the PV modules can be expected to experience (see Chapter 4)

Source: Global Sustainable Energy Solutions

The maximum input voltage is the maximum DC voltage that the inverter is designed to handle safely. If the array's open-circuit voltage exceeds the maximum input voltage, it may damage the inverter's electronics.

The vast majority of grid interactive inverters also have a maximum power point tracking (MPPT) range with specified minimum and maximum voltages. Within this range, the inverter tracks the maximum power point to ensure the array performs as well as possible; outside this range the array is likely to underperform. The upper limit of the MPPT range may be the same as the maximum input voltage, or it may be specified as a lower voltage (i.e. the inverter can handle a higher voltage than the maximum MPPT range voltage, but will not track the MPP of the array to this higher limit).

Box 9.1 Different terminology

Manufacturers do not always use the same terminology. Depending on the manufacturer or reference material, the MPPT range may also be specified as MPP range, inverter operating window, operating voltage window or similar.

Inverter manufacturers may also quote a maximum input voltage, which can also be specified as maximum PV open-circuit voltage, maximum input voltage, maximum DC voltage or similar. If a maximum input voltage is not specified it is typically assumed to be the upper limit of the MPPT range; however, the manufacturer should be contacted to verify this. This book uses the terms 'MPPT range' and 'maximum input voltage' to describe these terms.

As discussed in Chapter 4, a typical array consists of several modules connected in series to form a string or several strings connected in parallel to form an array. Assuming all modules are identical, the output voltage of the array is that of a single string. The voltage of a string is determined by the number of modules in that string and this section demonstrates how to calculate the maximum number of modules allowed in a string. It is important that the output voltage of the string is below the maximum system voltage and below the maximum inverter input voltage; it is also desirable for it to be within the MPPT range – these characteristics are quoted on the inverter's data sheet. The importance of the MPPT range is understood but the inverter's minimum and maximum input voltages are of greater importance for the system designer as they will impact on the system's performance (minimum input voltage) and the inverter's safety (maximum input voltage).

The first step is to acquire the maximum and minimum temperatures at the installation site, known as the maximum and minimum ambient temperatures; from these the PV module cell temperatures can be calculated. The PV cells will operate at a much higher temperature than the air temperature quoted in weather reports. National codes should be consulted first as they may specify the ambient or cell temperatures that should be used in sizing calculations. The examples cited below use temperature data from Berlin, Germany, and Sydney, Australia. These cities have been chosen to illustrate the difference in sizing requirements between hot and cold climates. In Australia, national standards

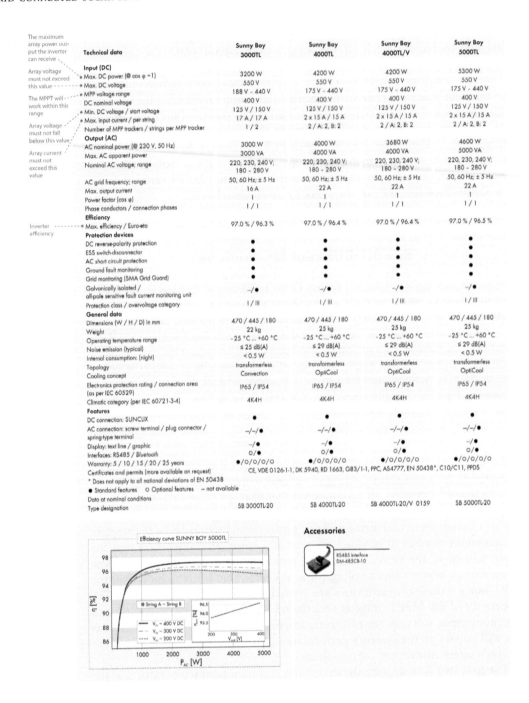

Technical data	Sunny Boy 3000TL	Sunny Boy 4000TL	Sunny Boy 4000TL/V	Sunny Boy 5000TL
Input (DC)				
• Max. DC power (@ cos φ =1)	3200 W	4200 W	4200 W	5300 W
• Max. DC voltage	550 V	550 V	550 V	550 V
• MPP voltage range	188 V – 440 V	175 V – 440 V	175 V – 440 V	175 V – 440 V
DC nominal voltage	400 V	400 V	400 V	400 V
• Min. DC voltage / start voltage	125 V / 150 V	125 V / 150 V	125 V / 150 V	125 V / 150 V
• Max. input current / per string	17 A / 17 A	2 x 15 A / 15 A	2 x 15 A / 15 A	2 x 15 A / 15 A
Number of MPP trackers / strings per MPP tracker	1 / 2	2 / A: 2, B: 2	2 / A: 2, B: 2	2 / A: 2, B: 2
Output (AC)				
AC nominal power (@ 230 V, 50 Hz)	3000 W	4000 W	3680 W	4600 W
Max. AC apparent power	3000 VA	4000 VA	4000 VA	5000 VA
Nominal AC voltage; range	220, 230, 240 V; 180 – 280 V	220, 230, 240 V; 180 – 280 V	220, 230, 240 V; 180 – 280 V	220, 230, 240 V; 180 – 280 V
AC grid frequency; range	50, 60 Hz; ± 5 Hz	50, 60 Hz; ± 5 Hz	50, 60 Hz; ± 5 Hz	50, 60 Hz; ± 5 Hz
Max. output current	16 A	22 A	22 A	22 A
Power factor (cos φ)	1	1	1	1
Phase conductors / connection phases	1 / 1	1 / 1	1 / 1	1 / 1
Efficiency				
Max. efficiency / Euro-eta	97.0 % / 96.3 %	97.0 % / 96.4 %	97.0 % / 96.4 %	97.0 % / 96.5 %
Protection devices				
DC reverse-polarity protection	●	●	●	●
ESS switch-disconnector	●	●	●	●
AC short circuit protection	●	●	●	●
Ground fault monitoring	●	●	●	●
Grid monitoring (SMA Grid Guard)	●	●	●	●
Galvanically isolated / all-pole sensitive fault current monitoring unit	–/●	–/●	–/●	–/●
Protection class / overvoltage category	I / III	I / III	I / III	I / III
General data				
Dimensions (W / H / D) in mm	470 / 445 / 180	470 / 445 / 180	470 / 445 / 180	470 / 445 / 180
Weight	22 kg	25 kg	25 kg	25 kg
Operating temperature range	–25 °C ... +60 °C	–25 °C ... +60 °C	–25 °C ... +60 °C	–25 °C ... +60 °C
Noise emission (typical)	≤ 25 dB(A)	≤ 29 dB(A)	≤ 29 dB(A)	≤ 29 dB(A)
Internal consumption: (night)	< 0.5 W	< 0.5 W	< 0.5 W	< 0.5 W
Topology	transformerless	transformerless	transformerless	transformerless
Cooling concept	Convection	OptiCool	OptiCool	OptiCool
Electronics protection rating / connection area (as per IEC 60529)	IP65 / IP54	IP65 / IP54	IP65 / IP54	IP65 / IP54
Climatic category (per IEC 60721-3-4)	4K4H	4K4H	4K4H	4K4H
Features				
DC connection: SUNCLIX	●	●	●	●
AC connection: screw terminal / plug connector / spring-type terminal	–/–/●	–/–/●	–/–/●	–/–/●
Display: text line / graphic	–/●	–/●	–/●	–/●
Interfaces: RS485 / Bluetooth	o/●	o/●	o/●	o/●
Warranty: 5 / 10 / 15 / 20 / 25 years	●/o/o/o/o	●/o/o/o/o	●/o/o/o/o	●/o/o/o/o
Certificates and permits (more available on request)	CE, VDE 0126-1-1, DK 5940, RD 1663, G83/1-1, PPC, AS4777, EN 50438*, C10/C11, PPDS			
* Does not apply to all national deviations of EN 50438				
● Standard features o Optional features – not available				
Data at nominal conditions				
Type designation	SB 3000TL-20	SB 4000TL-20	SB 4000TL-20/V 0159	SB 5000TL-20

The maximum array power output that the inverter can receive

Array voltage must not exceed this value

The MPPT will work within this range

Array voltage must not fall below this value

Array current must not exceed this value

Inverter efficiency

Accessories

RS485 interface DM-485CB-10

Efficiency curve SUNNY BOY 5000TL

η [%]

String A – String B

Vₙ = 400 V DC
Vₙ = 300 V DC
Vₙ = 200 V DC

P_AC [W]

www.SMA-Solar.com

SMA Solar Technology AG

Figure 9.4 The inverter features required for the sizing methodology used in this chapter can be found on the inverter's data sheet

Source: SMA Solar Technology AG

dictate that the maximum operating cell temperature is 25°C above the ambient temperature. This figure is often used in the US as well, although it is not standardized. The minimum operating cell temperature is the same as the minimum ambient temperature since the PV module will not have heated up when it starts generating electricity early in the morning.

Example 1: Sydney, Australia

In Sydney the ambient temperature can vary from 0°C to 50°C (32°F to 122°F), so the maximum and minimum cell temperatures are as follows:

Ambient temperature 0°C =	Minimum cell temperature = Minimum ambient temperature	= 0°C
Ambient temperature 50°C =	Maximum cell temperature = 50°C + 25°C	= 75°C

Example 2: Berlin, Germany

In Berlin the ambient temperature can vary from –10°C to 30°C (14°F to 86°F), so the maximum and minimum cell temperatures are as follows:

Ambient temperature of –10°C =	Minimum cell temperature = minimum ambient temperature	= –10°C
Ambient temperature of 40°C =	Maximum cell temperature = 40°C + 25°C	= 65°C

As discussed in Chapter 4 the temperature of the PV array affects its performance. The next step is to work out how the expected cell temperatures as calculated above will affect the voltage output of the array. To do this, information from the PV module's data sheet is required. This PV data will generally provide at least one voltage temperature coefficient, a specific rating that describes the effect of temperature on the cell voltage. This is usually expressed as a percentage or in volts per degree, whereas some manufacturers only provide a P_{max} (maximum power) temperature coefficient that can be used as an approximation for the voltage temperature coefficient.

Using the voltage temperature coefficient data, the PV array's maximum and minimum output voltages over the year can be calculated to ensure that they sit within the required operating voltage range for the inverter. The maximum power point voltage (V_{mp}) of the array should not fall below the minimum operating voltage of the inverter. If the module string voltage falls below the minimum operating voltage of the inverter, the inverter will either turn off (so the array will produce no power) or it will not operate at the array's maximum possible output.

The PV modules used in these examples are monocrystalline silicon modules produced by Sharp and the inverter is an SMA Sunny Boy 3000.

In order to calculate the module voltage at a particular temperature, one of the following formulas can be used:

If the temperature is higher than 25°C:

$$Voltage_{at\ X°C} = Voltage_{at\ STC} - [\gamma_V \times (T_{X°C} - T_{STC})]$$

Electrical Characteristics	
Cell	48 Monocrystalline (155.55mm)² Sharp silicon solar cells
No. of Cells and Connections	48 in series
Open Circuit Voltage (V_{oc})	30.2V
Maximum Power Voltage (V_{pm})	24V
Short Circuit Current (I_{sc})	8.54A
Maximum Power Current (I_{pm})	7.71A
Maximum Power (pm)[1]	Typical 185W
Encapsulated Solar Cell Efficiency (ηc)	15.9%
Module Efficiency (ηm)	14.1%
Maximum System Voltage	DC 1000V
Series Fuse Rating	15A
Type of Output Terminal	Lead Wire with MC3 Connector

[1](STC) Standard Test Conditions: 25°C, 1kW/m², AM 1.5

Figure 9.5 Electrical characteristics from module data sheet

Source: Sharp

Temperature Coefficient		
Temp. Coefficient of P_{max}	−0.485	% / °C
Temp. Coefficient of V_{oc}	−0.104	V / °C
Temp. Coefficient of I_{sc}	0.053	% / °C

Figure 9.6 Temperature coefficients from module data sheet

Source: Sharp

or

If the temperature is lower than 25°C:

$$Voltage_{\,at\,X°C} = Voltage_{at\,STC} + [\gamma_V \times (T_{X°C} - T_{STC})]$$

where:

$Voltage_{\,at\,X°C}$ = voltage at the specified temperature (X°C) in volts;

$Voltage_{\,at\,STC}$ = voltage at STC, i.e. the rated voltage in volts;

γ_V = voltage temperature coefficient in V/°C (absolute value);

$T_{X°C}$ = cell temperature in °C;

T_{STC} = temperature at standard test conditions (i.e. 25°C) in °C.

Note: The second formula as stated is not strictly correct but is a simplification of the concepts; however, it produces the correct answer for the operating conditions specified.

Example 1: Sydney, Australia

Ambient temperature of 0°C =	Corresponding module cell temperature = 0°C	= 0°C
Ambient temperature of 50°C =	Corresponding module cell temperature = 50°C + 25°C	= 75°C

Calculating maximum voltage

The module's maximum voltage (V_{oc}) is present at the minimum cell temperature which in this case is 0°C; it is therefore important to use the open-circuit voltage (V_{oc}) and adjust this figure according to the temperature coefficient when calculating the maximum voltage. From the data sheet it can be seen that the open-circuit voltage (V_{oc}) at standard test conditions is 30.2V, therefore:

The difference between the cell temperature and 25°C is calculated	0°C −25°C = −25°C
This figure is then multiplied by the temperature coefficient to calculate the increase in voltage	−25°C × −0.104V/°C = 2.60V
Finally the maximum voltage can be calculated	30.2V + 2.60V = 32.80V

Calculating minimum voltage

The module's minimum voltage will occur when the cell is hottest, i.e. at a cell temperature of 75°C. This figure is calculated using the maximum power voltage (V_{pm} or V_{mp}) and corresponding temperature coefficient. A temperature coefficient is not given for maximum power voltage, so the temperature coefficient for maximum power should be used as an approximation – this is given in per cent/°C on the data sheet and so must be converted to V/°C:

Temperature Coefficient		
Temp. Coefficient of P_{max}	−0.485	% / °C
Temp. Coefficient of V_{oc}	−0.104	V / °C
Temp. Coefficient of I_{sc}	0.053	% / °C

Figure 9.7 Temperature coefficients from the Sharp module data sheet as shown previously

Source: Sharp

The temperature coefficient should be converted into a decimal	$-0.485\% = -0.00485$
The temperature coefficient (V) per °C is calculated for module V_{pm} or V_{mp}	$-0.00485 \times 24V =$ $-0.1164V/°C$
Use this information to calculate the minimum voltage as follows:	
Calculate the difference between the cell temperature and STC	$75°C - 25°C = 50°C$
Multiply this by the V_{mp} or V_{pm} temperature coefficient	$50°C \times -0.1164V/°C$ $= -5.82V$
The voltage de-rating is subtracted from the $V_{mp} =$ minimum array voltage	$24V - 5.82 V =$ $18.18V$

Thus the minimum module voltage will be 18.18V. The maximum voltage (i.e. 32.8V) and minimum voltage (i.e. 18.18V) are used to calculate how many modules are allowed in each string. It is also common practice to include a safety margin in each of these figures, normally 10 per cent above the specified minimum voltage and 5 per cent below the maximum input voltage.

Calculating the minimum number of modules in a string

An outline of voltage drop calculations is given in Chapter 10.

The expected voltage drop across the DC cables should be included in this calculation: 1% voltage drop is assumed, so the minimum voltage should be decreased by the assumed 1% voltage loss (therefore the voltage figure is multiplied by 0.99)	$18.18V \times 0.99 =$ $17.99V$
The minimum inverter input voltage (from data sheet) should be increased by the 10% safety margin required (therefore the voltage figure is multiplied by 1.1)	$268V \times 1.1 = 294.8 V$
Finally, the minimum number of modules in the string is calculated by dividing this figure by the minimum module voltage	$294.8V/17.99V =$ 16.39 modules
At least 17 modules must be connected in series to form a string	

Table 9.1 SMA Sunny Boy 3000 technical data

Technical data	Sunny Boy 3000
Max DC power	3200W
Max DC voltage	600V
MPP voltage range	268V–480V
DC nominal voltage	350V
Min DC voltage/start voltage	268V/330V

Source: SMA-Australia

Calculating the maximum number of modules in a string

First the maximum inverter input voltage is reduced to account for the 5% safety margin (therefore the maximum voltage figure is multiplied by 0.95)	600V × 0.95 = 570V
The maximum inverter voltage should be divided by the maximum module voltage (as previously calculated to determine the maximum allowable modules in a string)	570V/32.8V = 17.38 modules

This figure should always be rounded down for safety reasons, so each string must have 17 modules.

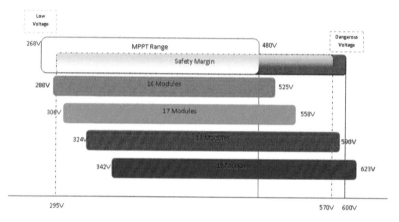

Figure 9.8 Four possible string lengths using the Sharp modules. It can be seen that if there are 16 or fewer modules the voltage range of the string falls below the safety margin and the inverter will turn off on very hot days when the voltage drops below the MPPT range. If there are 18 or more modules the voltage will exceed the inverter's maximum DC input voltage on cold days, damaging the inverter

Source: Global Sustainable Energy Solutions

Example 2: Berlin, Germany

Using the same methodology for a location with a very different temperature range (–10°C to 40°C), we will now do the same calculation. The maximum voltage will occur at the minimum cell temperature of 10°C and can be calculated using the open-circuit voltage temperature coefficient (–0.104V/°C) as follows:

Calculating the maximum voltage

The difference between the cell temperature and 25°C is calculated	−10°C −25°C = −35°C
This figure is then multiplied by the temperature coefficient to calculate the increase in voltage	−35°C × −0.104V/°C = 3.64V
Finally the maximum voltage can be calculated	30.2V + 3.64V = 33.84V

Calculating the maximum number of modules in a string

First the maximum inverter input voltage is multiplied by 0.95 to account for the 5% safety margin	$600V \times 0.95 = 570V$
The maximum inverter voltage should be divided by the maximum module voltage (as previously calculated to determine the maximum allowable modules in a string)	$570V/33.84V = 16.84$ modules

This figure should always be rounded down for safety and so the string can only have 16 modules.

Calculating the minimum voltage

Minimum voltage will occur at a cell temperature of 65°C and is calculated as follows:	
The temperature coefficient (V) per °C is calculated for module V_{pm}/V_{mp}	$-0.00485 \times 24V = -0.1164V/°C$
Calculate the difference between the cell temperature and STC	$65°C - 25°C = 40°C$
Then multiply this by the V_{mp} temperature coefficient	$40°C \times -0.1164V/°C = -4.66V$
Take this last figure away from the rated V_{mp} (minimum voltage figure is calculated for maximum temperature conditions)	$24V - 4.66V = 19.34V$

Calculating the minimum number of modules in a string

The minimum module voltage is multiplied by the voltage drop expected across the DC cables (1%)	$19.34V \times 0.99 = 19.15V$
The minimum inverter input voltage should be multiplied by 1.1 to account for the 10% safety margin required	$268V \times 1.1 = 294.8$ V
Finally, the minimum number of modules in the string is calculated by dividing this figure by the minimum module voltage	$294.8V/19.15V = 15.39$

This figure should be rounded up (because it a minimum) so at least 16 of these modules must be connected in series to form a string.

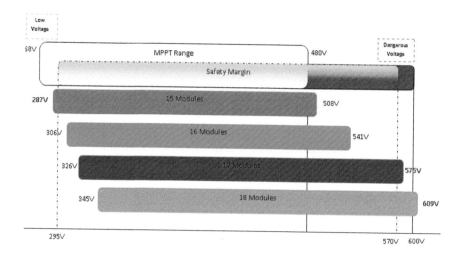

Figure 9.9 Each string must be 16 modules

Source: Global Sustainable Energy Solutions

Matching current specifications

Inverter manufacturers generally give the following ratings in terms of current:

- Maximum DC input current: maximum DC current that the inverter can process;
- Maximum AC output current: maximum AC current that the inverter can deliver.

The current of a PV module does not vary as dramatically as the voltage does. However, it does increase slightly as the temperature increases. Because this increase is so small, PV array designers often do not account for it when sizing the array to match the inverter, unless the array's current output is very close to the inverter's maximum current input.

It is important to remember that an array is normally several strings connected in parallel and the input DC current will be the sum of the currents of all the strings connected to the inverter. The short-circuit current is used in these calculations because it is the largest current the module will produce.

Examples: Both examples have the same number of strings as temperature effects are not being considered:

12A/8.54A = 1.41 strings. This is rounded down to 1 string.

Table 9.2 SMA Sunny Boy 3000 technical data

Technical data	Sunny Boy 3000
Max input current/per string	12A/12A
Number of MPP trackers/strings per MPP tracker	1/3
Max output current	15A

Source: SMA Solar Technology AG

Matching modules to the inverter's power rating

Most inverter manufacturers give a number of ratings for their inverters in terms of power. Common ratings include:

- Maximum PV array rated power: Maximum rated power of the PV array, usually in kWp or Wp;
- Maximum DC input power: Maximum amount of DC power that the inverter can convert to AC (this is generally lower than the maximum PV array power, because of losses in PV arrays);
- Maximum AC output power: Maximum amount of AC power that the inverter can produce.

For most inverters, the maximum allowable power (watts-peak) of the array is greater than the maximum AC output of the inverter. The reasons for this are losses in the system (i.e. the array will not produce its rated power), as discussed later.

To calculate the total number of modules that can be connected to the inverter, the inverter's maximum PV array rated power is divided by the rated power of the module. If the inverter's maximum PV array rated power is not given, then the maximum DC input power should be used. In some cases it may not be possible to connect the maximum number of modules to the inverter as the voltage or current characteristics may not allow it. An example of the inverter's maximum array power figure quoted is shown in Figure 9.10.

Example: The maximum DC power of the inverter is 3200W, while the maximum power of the module is 185W, so 3200W/185W = 17.3.

The maximum number of modules that can be connected to the inverter is 17.

Example 1: Sydney, Australia

Given that the voltage specifications of the inverter limit the string to 17 modules and the power specification of the inverter limits the whole system to

Technical data	Sunny Boy 3000TL	Sunny Boy 4000TL	Sunny Boy 4000TL/V	Sunny Boy 5000TL
Input (DC)				
Max. DC power (@ cos φ =1)	3200 W	4200 W	4200 W	5300 W
Max. DC voltage	550 V	550 V	550 V	550 V
MPP voltage range	188 V – 440 V	175 V – 440 V	175 V – 440 V	175 V – 440 V
DC nominal voltage	400 V	400 V	400 V	400 V
Min. DC voltage / start voltage	125 V / 150 V	125 V / 150 V	125 V / 150 V	125 V / 150 V
Max. input current / per string	17 A / 17 A	2 x 15 A / 15 A	2 x 15 A / 15 A	2 x 15 A / 15 A
Number of MPP trackers / strings per MPP tracker	1 / 2	2 / A: 2, B: 2	2 / A: 2, B: 2	2 / A: 2, B: 2
Output (AC)				
AC nominal power (@ 230 V, 50 Hz)	3000 W	4000 W	3680 W	4600 W
Max. AC apparent power	3000 VA	4000 VA	4000 VA	5000 VA
Nominal AC voltage; range	220, 230, 240 V; 180 - 280 V	220, 230, 240 V; 180 - 280 V	220, 230, 240 V; 180 - 280 V	220, 230, 240 V; 180 - 280 V
AC grid frequency; range	50, 60 Hz; ± 5 Hz	50, 60 Hz; ± 5 Hz	50, 60 Hz; ± 5 Hz	50, 60 Hz; ± 5 Hz
Max. output current	16 A	22 A	22 A	22 A
Power factor (cos φ)	1	1	1	1
Phase conductors / connection phases	1 / 1	1 / 1	1 / 1	1 / 1
Efficiency				
Max. efficiency / Euro-eta	97.0 % / 96.3 %	97.0 % / 96.4 %	97.0 % / 96.4 %	97.0 % / 96.5 %
Protection devices				

Annotations (left margin):
- The maximum array power output the inverter can receive
- Array voltage must not exceed this value → Max. DC voltage
- The MPPT will work within this range → MPP voltage range
- Array voltage must not fall below this value → Min. DC voltage / start voltage
- Array current must not exceed this value → Max. input current / per string
- Inverter efficiency → Max. efficiency / Euro-eta

Figure 9.10 Extract from SMA inverter data sheet shown in Figure 9.4

Source: SMA Solar Technology AG

17 modules, the only option for this array is one string of 17 modules. The overall rated power is calculated as $17 \times 185\text{Wp} = 3145\text{Wp}$.

Example 2: Berlin, Germany

Two configurations are possible for the German system, one string of 16 modules. The overall rated power is calculated as $16 \times 185\text{Wp} = 2960\text{Wp}$.

Calculations need to be done for voltage, current and power when matching a module to an inverter. If the rated power of the array is much less than the inverter's, then the inverter's efficiency may decrease.

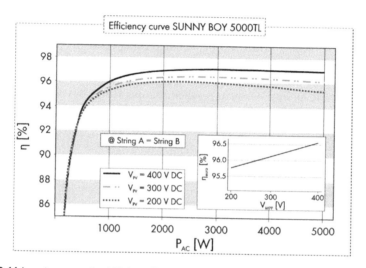

Figure 9.11 Inverters operate at higher efficiencies when they are connected to an array which produces the output they are designed to take. The inverter's efficiency increases against the voltage figures shown

Source: SMA Solar Technology AG

Box 9.2 What happens if more modules are connected to the inverter than it is rated for?

Sometimes designers install more modules than the inverter is rated for (i.e. installing 18 for the examples in this chapter), so that the array has the appropriate number of modules to fit voltage calculations. Another reason to add extra modules is because system designers understand that there will be unavoidable losses (dirt, temperature, cable losses etc.) between the array and the inverter, so the full rated power of the array is unlikely to reach the inverter. Designers may therefore choose to install a slightly smaller (and generally cheaper) inverter to a larger array (known as undersizing the inverter). In this case, the system designer must ensure that the system de-rates as outlined in the following section have been properly calculated and the correct information communicated to the client.

Losses in utility-interactive PV systems

There are a vast number of factors that prevent PV systems from working at maximum efficiency and reaching their rated output. These factors are accounted for through a process called de-rating. Calculation of de-rating factors for the major sources of loss is outlined here and the next section demonstrates how these factors are used in system yield calculations.

Temperature of the PV module

As already discussed (mainly in Chapter 4), the temperature at which the PV module operates has a large effect on its power output and high operating temperatures lead to power losses. The effect of temperature on output varies from module to module and can be calculated using the temperature coefficients provided on the manufacturer's data sheets. In order to make a more realistic estimation of system output it is useful to calculate the de-rating factor which indicates the resultant efficiency when temperature has been taken into account.

Example 1: Sydney, Australia

The de-rating factor for the Sharp modules (specifications given on page 131) installed in Sydney (average ambient temperature 23°C (73.4°F)) is calculated as follows:

From the ambient temperature the cell temperature is calculated	$23°C + 25°C = 48°C$
The difference between the cell temperature and the standard testing conditions (STC, 25°C) is then be calculated	$48°C - 25°C = 23°C$
If the temperature coefficient is given in %/°C it must be converted to a decimal	$0.485\%/°C = 0.00485$
Temperature coefficient is multiplied by the difference between cell temperature and STC	$0.00485 \times 23 = 0.11155$
To calculate f_{temp} this figure must be subtracted from 1	$f_{temp} = 1 - 0.11155 = 0.88845$
Therefore the resultant temperature efficiency is 88.8%	

Example 2: Berlin, Germany

The de-rating factor for the Sharp modules installed in Berlin (average ambient temperature 9.8°C (49.64°F)) is calculated as follows:

From the ambient temperature the cell temperature is calculated	$9.8°C + 25°C = 34.8°C$
The difference between the cell temperature and the standard testing conditions (STC 25°C)	$34.8°C - 25°C = 9.8°C$
The temperature coefficient is given in %/°C and converted to a decimal	$0.485\%/°C = 0.00485$

This number is then multiplied by the difference between cell temperature and STC	$0.00485 \times 9.8 = 0.04753$
To calculate f_{temp} this figure must be subtracted from 1	$f_{temp} = 1 - 0.04753 = 0.95247$
Therefore the resultant temperature efficiency is 95.25%	

From these two examples it can be seen that the temperature de-rating varies significantly depending on location (examples given: for Sydney 88.8 per cent and for Berlin 95.25 per cent) and that colder climates will experience fewer temperature-related losses. However, it is important to note that PV systems in cold climates will not necessarily have a greater power output as warmer climates are characterized by good solar irradiation while colder climates tend to experience shorter days and less sunshine.

Dirt and soiling

Losses caused by the build-up of dirt on the modules are often known as 'soiling' losses. If dirt and debris (such as leaves) build up on the surface of a PV module, they will shade the module and reduce its power output. The reduction in power output will depend largely on the site and the factors influencing soiling such as the tilt angle at which the modules are installed (when modules are installed flat it is very difficult for the rain to wash away any accumulated dirt), the dust levels on site, local pollution that may form a film on the glass, the frequency of rainfall and whether or not the environment is salty, i.e. close to the coast. The de-rating factor due to soiling can only be estimated. It is fairly common practice to use $f_{dirt} = 0.90$ (10 per cent output reduction due to soiling) for very dirty sites and $f_{dirt} = 0.95$ (5 per cent output reduction due to soiling) for relatively clean sites that experience regular rainfall. See Chapter 12 for soiling losses which can be reduced by cleaning the modules.

Manufacturer's tolerance

The manufacturer's tolerance accounts for the small output variations between individual modules produced by the same manufacturer and is normally specified on the module data sheet. If the manufacturer's tolerance shows a negative loss, i.e. –3 per cent, the peak power of the module may be 3 per cent less than given on the data sheet, i.e. the minimum peak power from the 315 watt modules produced by that manufacturer is 305.55W (i.e. 315W minus 3 per cent). Some manufacturers also give a positive manufacturer's tolerance, i.e. +5 per cent meaning some of the modules produced will have a higher peak power than given on the data sheet (i.e. 315W plus 5 per cent = 330.75W for the example in Figure 9.12).

Shading

Shading can be a major contributor to loss of system performance, for two main reasons:

Electrical Data		
Measured at Standard Test Conditions (STC): irradiance of 1000/m², AM 1.5 and cell temperature 25°C		
Peak Power (+5/-3%)	P_{MAX}	315 W
Rated Voltage	V_{MP}	54.7 V
Rated Current	I_{MP}	5.76 A
Open Circuit Voltage	V_{OC}	64.6 V
Short Circuit Current	I_{SC}	6.14 A
Maximum System Voltage	UL	600 V
Temperature Coefficients		
	Power	-0.38% / K
	Voltage (V_{OC})	-176.6mV / K
	Current (I_{SC})	3.5 mA / K
NOCT		45°C +/- 2°C
Series Fuse Rating		15 A

Figure 9.12 Sample data sheet showing manufacturer's tolerance

Source: Global Sustainable Energy Solutions

- Shading leads to a voltage drop and if enough of the PV array is shaded the voltage may drop outside the inverter's voltage window causing the inverter to turn off and the system will produce no power.
- Where multiple strings are connected in parallel, the shading on one string may impact on the others. This is because current and voltage are directly affected by the irradiation received (Chapter 4) and when the module is shaded, it does not receive as much irradiation.

If a string is shaded it will generate a lower voltage than the other (unshaded) strings. As discussed in Chapter 4, the voltage generated by the unshaded strings connected in parallel will also be reduced so that all strings generate the same voltage. One way to avoid this problem is to use a multi-string inverter (see Chapter 5).

The best way to avoid these issues is to install the array in an area that does not experience any shading. Some experienced PV array designers may be able to design systems around the problem of partial shading, but in general this practice is to be avoided or only used by experienced professionals.

Orientation and module tilt angle

The losses are caused by the orientation and tilt angle at which an array is installed (see Chapter 7). In practice calculating the effect of these angles on energy yield is done with tables based on experimental data; reference tools are given in Chapter 15.

Voltage drop

The calculation of voltage drop is discussed in Chapter 10. In general, designers try to minimize voltage drop by using larger cables as the cost of buying a large cable is significantly less than the cost of adding additional modules to the array to compensate for power loss: for example a 5 per cent voltage drop in a 3000Wp array is equivalent to the power output of one 105Wp PV module (voltage drop and power loss are directly related because P = IV). The maximum voltage drop allowed will be specified by the codes: in Australia it is 5 per cent, but the industry standard and many designers aim for voltage drops closer to 1 per cent for reasons explained. The US National Electric Code (NEC) has no absolute requirement for voltage drop, but the North American Board of Certified Energy Practitioners (NABCEP) recommends that the overall system voltage drop be limited to between 2 and 5 per cent of the circuit operating voltage and that the voltage drop in DC conductors be below 1 per cent. As well as by increasing cable size, voltage drop can also be decreased by carefully planning and minimizing the length of cable runs. When designing a system for an area with a feed-in tariff, any losses attributable to the system's voltage drop can be quantified and costed to see if it is financially advantageous to increase the cable sizing to avoid these power losses.

Inverter efficiency

Power losses occur within the inverter because of heat losses from the electronics and transformer (inverters that are transformer-less therefore tend to have higher efficiencies). The losses are reflected by the inverter efficiency given in the manufacturer's data sheet.

DC Input Data	Inverter 2000	Inverter 3000	Inverter 2500-LV
Recommended PV power	1500 – 2500 W$_p$	2500 – 3300 W$_p$	1800 – 3000 W$_p$
Max. DC input voltage	500 V	500 V	500 V
Operating DC voltage range	150 – 450 V	150 – 450 V	150 – 450 V
Max. Usable DC input current	13.6 A	18 A	16.9 A
AC Output Data	**Inverter 2000**	**Inverter 3000**	**Inverter 2500-LV**
Maximum output power @ 40°C	2000 W	2700 W	2350 W
Nominal AC output voltage	240 V		208 V
Utility AC voltage range	212 – 264 V (240 V +10% / -12%)		183 – 229 V
Maximum AC current	8.35 A	11.25 A	11.25 A
Maximum utility back feed current	0.0 A	0.0 A	0.0 A
Operating frequency range	59.3 – 60.5 Hz (60 Hz nom)		
Total Harmonic Distortion THD	< 5%		
Power Factor (cos phi)	1		
General Data	**Inverter 2000**	**Inverter 3000**	**Inverter 2500-LV**
Max. Efficiency	95.2%	95.2%	94.4%
Consumption in stand-by	< 0.15 W (night)		
Consumption during operation	7 W		
Enclosure	NEMA 3R		
Size (l x w x h)	18.5 x 16.5 x 8.85 inches (470 x 418 x 223 MM)		
Weight	26 lbs. (11.8 kg)		
Ambient temperature	-5 to 122°F (-20 to +50°C)		
Cooling	Controlled forced ventilation		
Integrated AC and DC disconnects	Standard UL approved DC and AC disconnects		

Figure 9.13 Sample data sheet showing data for three different inverters and their efficiencies

Source: Global Sustainable Energy Solutions

The inverter efficiency will also vary depending on the actual input power, as previously discussed, and the operating temperature. It is important that the inverter is installed in a cool, well-ventilated area and never installed in direct sunlight. The ventilation requirements of the inverter are given in the manufacturer's installation recommendations.

Calculating system yield

The PV system owner is generally most interested in the final system yield and the installer should provide them with an estimate that takes into consideration the size of the PV array, the amount of irradiation it will receive and the system losses.

Example 1: Sydney, Australia

The size of the PV array was determined earlier in this chapter to be one string of 17 modules, so the PV array's rated power is 3145Wp.

The amount of irradiation received at the site was calculated by consulting the Australian Bureau of Meteorology; on average Sydney receives 4.5 peak sun hours per day and this figure is multiplied by 365 to determine the annual irradiation (1643kWh/m²/year).

The system losses must also be accounted for by using the de-rating factors outlined above. The array is to be installed flat on a roof using standoff mounts and so must follow the orientation and tilt of the roof, which are 20° from true north and 40° respectively. The resulting system efficiency is determined from tables available on the Australian Clean Energy Council's website (www. cleanenergycouncil.org.au).

The total de-rating factor due to system losses is calculated by multiplying all the de-rating factors together, hence: $0.88845 \times 0.9 \times 1 \times 1 \times 0.98 \times 0.95 \times 0.95 = 0.71$, meaning for the conditions specified, the system will be 71 per cent efficient and system losses will be 29 per cent.

From this figure, the total system yield can be calculated:
1643kWh/m²/year $\times 3.145$kW$_p \times 0.71 = 3669$kWh/year
Hence the total system yield is 3669kWh per year.

ANNUAL DAILY IRRADIATION ON AN INCLINED PLANE EXPRESSED AS % OF MAXIMUM VALUE FOR SYDNEY									
	Plane Inclination (degrees)								
Plane Azimuth (degrees)	0	10	20	30	40	50	60	70	80	90
0	87%	94%	98%	100%	99%	97%	91%	84%	75%	64%
10	87%	94%	98%	100%	99%	96%	91%	84%	75%	64%
20	87%	93%	97%	99%	98%	95%	90%	83%	74%	64%
30	87%	93%	96%	98%	97%	94%	88%	81%	73%	63%
40	87%	92%	95%	96%	95%	91%	86%	79%	71%	63%
50	87%	91%	94%	94%	92%	89%	84%	77%	69%	61%
60	87%	90%	92%	91%	89%	86%	80%	74%	67%	60%

Figure 9.14 Extract from *Grid-Connect PV Systems: System Design Guidelines for Accredited Designers*

Source: Clean Energy Council (Australia)

Table 9.3 System losses (Sydney, Australia)

Source of loss	De-rating factor	Description
Temperature	0.88845	As calculated in the last section
Dirt/soiling	0.9	Sydney is a coastal area so losses due to soiling are assumed to be ~10%
Manufacturer's tolerance	1.0	This was not given on the data sheet so cannot be accounted for
Shading	1.0	Assume the array is unshaded
Orientation and tilt angle	0.98	From tables
Voltage drop	0.95	Assume a voltage drop of 5%
Inverter efficiency	0.95	Inverter efficiency is given on the data sheet to be 95%

Example 2: Berlin, Germany

The Berlin system is one string of 16 modules, therefore the PV array's rated power is 2960Wp. The yield of this system can be calculated using the PVGIS online tool (http://re.jrc.ec.europa.eu/pvgis) which requires several inputs. The program requires the system losses (excluding orientation and tilt angle which are optimized and temperature which the PVGIS system will calculate). PVGIS will also calculate losses due to angular reflectance effects. Figure 9.15 shows the input into the PVGIS online tool.

The total de-rating factor due to system losses is $0.95 \times 1 \times 1 \times 0.95 \times 0.95$ = 0.8574, meaning for the conditions specified, the system will be ~86 per cent efficient.

The total system losses as a percentage are calculated by subtracting the de-rating factor from 1 and then multiplying the result by 100: $1 - 0.8574 = 0.1426$. Therefore the total system losses are 14.26 per cent.

Table 9.4 System losses (Berlin, Germany)

Source of loss	De-rating factor	Description
Temperature	0.918	As calculated by PVGIS
Angular reflectance effects	0.97	As calculated by PVGIS
Dirt/soiling	0.95	Berlin is an urban area so losses are assumed to be ~5%
Manufacturer's tolerance	1.0	This was not given on the data sheet so cannot be accounted for
Shading	1.0	Assume the array is unshaded
Voltage drop	0.95	Assume a voltage drop of 5%
Inverter efficiency	0.95	Inverter efficiency is given on data sheet as 95%

The temperature losses (not included in the above calculation of system losses) are calculated by PVGIS based on the location of the system and the variety of PV module. In this case the PVGIS tool calculates the temperature losses to be 8.2 per cent based on temperature data for Berlin and the variety of module (crystalline silicon in this case). This corresponds to a temperature de-rating factor of 0.918. This figure differs from the calculation on page 139 because two different methods for calculating temperature de-rating factor have been used. These methods are based on different assumptions, for example PVGIS uses data for a generic crystalline PV module whereas the calculation in this book uses data specific to the Sharp module being used.

The PVGIS tool shows that the irradiation incident on the module is approximately 1140kWh/m²/year. Using this figure, the total system yield can be calculated:

$$1140kWh/m^2/year \times 2.960kWp \times 0.7635 = 2576kWh/year$$

Total system efficiency is 76%. The total system yield is 2576kWh per year, which is approximately 70 per cent of the yield of the system installed in Sydney. The difference in irradiation between the two locations is the primary reason for the large difference in system yield.

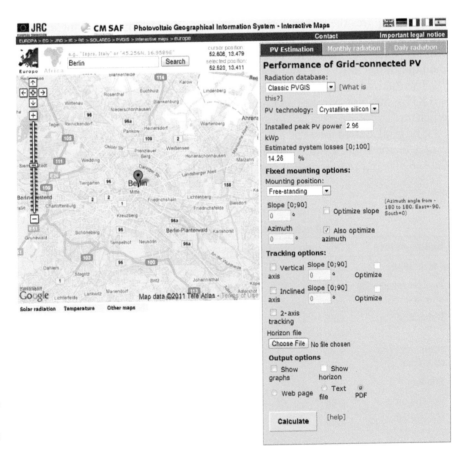

Figure 9.15 Input into PVGIS online tool. Details on how to use the tool are given on the website

Source: PVGIS © European Communities, 2001–2008

10
Installing Grid-connected PV Systems

Guidelines for the installation of PV systems vary by country, as there are different standards and regulations to apply in various regions. The specific permits, requirements, metering policies, interconnection standards, installer authorizations and other country-specific codes and regulations will determine the procedures undertaken when installing a PV system. The manufacturer's instructions should also be carefully followed during the installation. The main issues discussed below include installation of system components and requirements for connecting a PV system to the utility grid.

PV array installation

It is highly advisable that PV installers take time to carefully plan the precise location of the array before deciding on the installation methodology. This is commonly done by measuring the available installation area and marking the boundaries of the array(s) as well as the location of the mounting system's attachment points on the installation area, i.e. the roof, using a string or chalk line. The next step is to install the attachment hardware (array mounting structure) to secure the mounting system to the roof. If mounting on a rooftop, care should be taken to ensure that attachment screws are securely embedded into the rafters or other structural supporting members to provide maximum attachment strength.

Once the attachment points are secured, the mounting system should be assembled. Several types of mounting systems for PV modules are available and the choice depends on the specific application. The main types of mounting systems and their installation are explained in Chapter 6. Once the mounting system is fully assembled, the process of installing PV modules can begin. Many proprietary mounting systems rely on a compression clamp to secure the module frame to the mounting rails. It is imperative that the module clamps are fully compatible with the mounting system being installed.

DC wiring

Typically, PV modules intended for use in grid-connected systems are supplied complete with the interconnection cables wired from a sealed junction box with plug and receptacle connectors at the end of each cable length. A string of modules is formed by connecting adjacent modules together in series (positive to negative or negative to positive). Once the desired number of modules is connected in series and that string is formed, the individual circuit needs to be brought to a central location, typically a PV combiner box, where it will be connected in parallel with any additional series strings. Any module string fusing will also be installed in the PV combiner box. DC wiring is a very important component of the PV system and there are several key factors that must be considered during design and installation.

Cabling routes and required lengths

Once the location of all equipment has been decided, the cable route must be determined. The main cabling routes are shown in the Figure 10.1. For any installation, installers should seek to minimize the length of these cable routes. As discussed later in this chapter, the cable length and cross-section determines the voltage and power losses sustained.

Additionally when planning cable routes, the PV array wiring should be installed in such a way to minimize any conductive loops. Reducing conductive

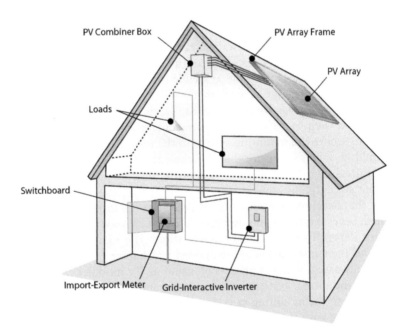

Figure 10.1 The main cabling routes are shown in red; if the installation does not include a PV combiner box then the cable will run directly from the PV array to the inverter. This diagram shows a net-metering arrangement; different metering arrangements are introduced in Chapter 5 and discussed later in this chapter

Source: Global Sustainable Energy Solutions

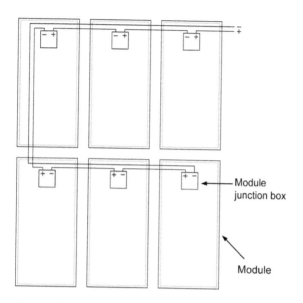

Figure 10.2 Example of correct PV wiring where conductive loops have been reduced

Source: Global Sustainable Energy Solutions

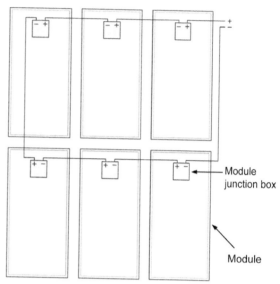

Figure 10.3 Example of incorrect PV wiring where the wiring is a conductive loop

Source: Global Sustainable Energy Solutions

Figure 10.4 Photograph of a conductive loop

Source: Global Sustainable Energy Solutions

loops will lower the risk of lightning induced over-voltages in the array wiring, as well as reducing interference to AM/FM radio signals.

The cable route from the PV array to the inverter (via the PV combiner box) is DC and a cable rated at the appropriate DC voltage and current should be used. Local regulations and country-specific standards will determine the required current-carrying capacity and voltage rating of cabling used. Copper conductors are generally preferred to aluminium and are available in single- or multi-strands. All cabling is insulated to protect the wire from the surrounding environment and to protect people and equipment. Cable insulation varies based on ratings for temperature, sunlight, oil or water resistance and location (dry, wet). The colour of the insulation indicates the polarity of the conductor and is generally governed by national electrical codes/standards. The standards for cable insulation colour differ greatly from one country to the next. In many parts of the world brown insulation indicates a positive conductor and blue or grey insulation indicates a negative conductor. However, red (positive) and black (negative) is also a very common colour combination. Green and yellow insulation typically indicates a grounded/earthed conductor, but this is not the case everywhere. It is imperative that the colouring given in local wiring codes is followed during installation.

Cable sizing

When selecting any cable for a system (i.e. sizing the cable) both current-carrying capacity (CCC) and the expected voltage drop must be considered, i.e. how much current the conductor is able to carry and the power losses sustained due to the conductor's internal resistance.

Cables are classified by their rated CCC at certain operating temperatures. The CCC indicates the amount of current that can pass through the wire or cable before it sustains damage. CCC is determined by: wire type (copper or aluminium), wire size/gauge, insulation rating (wet rating for outdoors), highest insulation temperature and location (free air, conduit or buried). CCC decreases with increasing temperature.

Voltage drop is a crucial issue in PV systems. Voltage drop increases with increasing current and decreasing conductor size (i.e. the smaller the cross-sectional area of the cable or wire, the higher the internal resistance of the conductor, hence the larger the voltage drop). The sizes and resistances of terminations, fuses and disconnect/isolator devices may also contribute to voltage drop. For grid-connected PV systems, it is common practice for cables to be sized so that there is no more than a 1 per cent voltage drop through either DC or AC conductors. Most national codes specify a maximum allowable voltage drop and should be consulted; in Germany, for instance, codes specify a maximum voltage drop of 1 per cent, but in Australia a 5 per cent voltage drop is currently permissible. As voltage drop represents a reduction in system power output, reducing voltage drop in cabling as much as possible is recommended. It is often cheaper to purchase larger cable than to purchase additional modules to compensate for the loss of output. If voltage drop is not properly accounted for, it may affect the operation of the inverter sensing circuits.

Box 10.1 Voltage drop calculation

The length of cabling route is 15m from the PV array to inverter; copper cabling (resistivity of copper is $0.0183\Omega/m/mm^2$) is used with a cross-sectional area of $2.5mm^2$ and it must carry a current of 5A. According to Ohm's law, voltage drop is calculated as:

Voltage drop = 2 × length × current × resistance (2 accounts for 2 cables, 1 +ve and 1 −ve)

Here resistance = resistivity/area, so

Voltage drop = (2 × length [in metres] × current [in amps] × resistance [in $\Omega/m/mm^2$])/area [in mm^2]

Voltage drop = (2 × 15m × 5A × $0.0183\Omega/m/mm^2$)/$2.5mm^2$

Voltage drop = 1.098V

If the voltage at the maximum power is known, then the voltage drop as a percentage can be calculated; if the voltage at maximum power is known to be 155V then the voltage drop is calculated as follows:

Voltage drop (%) = Voltage drop/voltage at maximum power × 100%

Voltage drop (%) = 1.098V/155V × 100%

Voltage drop (%) = 0.71%

Therefore if this voltage drop is sustained in a $10kW_p$ installation, the power loss sustained will be 71Wp, meaning the installation has effectively been reduced to 9.929kWp PV installed before any other system de-ratings are applied (see Chapter 9 for system de-ratings).

US installers often deal with American wire gauge (AWG) cable sizes. It is necessary to convert these gauges into the equivalent mm^2 value for use in the above formula. Conversions are given in Table 10.1

Table 10.1 US wire gauge conversion table

AWG	Size in mm^2
14	2.00
12	3.31
10	6.68
8	8.37
6	13.30
4	21.15
2	33.62
1	42.41
0	53.50

PV array cables should also be rated for use in wet conditions, be UV stabilized, have suitable insulation for protection from the elements, and be rated for the maximum system output voltage and current in accordance with local electrical standards. Many companies are now manufacturing cable labelled 'solar cable' specifically for use in PV systems. Solar cable is designed to ensure safety in the outdoor environment, providing high resistance to UV and double insulation (also known as double sheathing). Solar cable is designed to carry DC current and voltage and many local standards require it to be marked so that it can easily be distinguished from other power cables. It is available in many sizes allowing system designers flexibility in ensuring that voltage drop is minimized. Solar cable is often flexible, so it is easier to install. It is good practice whenever possible for cables to be installed out of direct sunlight and secured so that they are not fixed directly on the roof and cannot move around in the wind. Mechanical protection greatly reduces the risk of cabling being damaged and possibly causing a ground/earth fault or even a fire. Cables should always be secured and conduit may be used to protect them. As well as deteriorating under adverse weather conditions, cables are also at risk of attack by animals such as possums or rodents; laying cable in conduit can provide protection from all these problems.

PV combiner box

PV combiner boxes (discussed in Chapter 5) are used to combine multiple PV strings into fewer parallel circuits in order to reduce the amount of wiring required. The PV combiner box enclosure must be the correct size, so it adequately houses the correct amount of cables without any risk of crushing them. The cables inside the combiner box should be appropriately coloured and labelled and the box must be rated suitable for use in the environment in which it is installed.

Figure 10.5 Example of an array cable left loose and unsupported

Source: Global Sustainable Energy Solutions

Figure 10.6 An electrician installs a small white PV combiner box next to the array

Source: Global Sustainable Energy Solutions

System grounding/earthing

Grounding, also referred to as earthing, is used to ensure that the exposed conductive parts (i.e. the array frame) of a system are equipotential, which means there is no voltage difference between the components and the ground. The metal frames and metal mounting structure of PV panels can be grounded so that the voltage in these surfaces does not reach dangerous levels. This ensures that a person touching a conductive component will not receive an electric shock. It is important to remember that people coming in contact with the PV array, i.e. the system owner cleaning the modules, may have very little experience with electrical systems and ensuring their safety is paramount. The grounding/earthing rules and standards vary widely between countries, and the system and individual components should be earthed as per the standard in that country. Certain module types will require earthing to ensure maximum performance of the modules (e.g. SunPower module range) and this is a requirement of the module manufacturer. The installer should ensure that the inverter and module manufacturers' instructions are followed as well as national codes. See Figure 10.7.

Inverter installation

The inverter should be installed as close as possible to the modules in order to minimize DC cable lengths (longer cables lead to larger power losses). Inverters should be located in a shady, sheltered and well-ventilated area. They should not be exposed to temperatures outside the range specified on their data sheet (normally −25°C to 60°C). Additionally, the wall on which the inverter is installed must be able to support its load. Over-current protection and

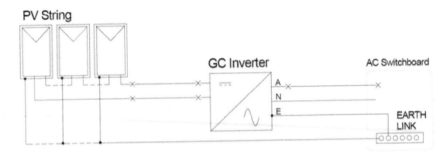

Figure 10.7 It is common practice to ground each component of the PV system separately so that if one is removed the others will remain grounded, i.e. the earth of the array is not connected to the earthing wire from the inverter. The technique shown in this diagram is compliant with Australian National Standards. Techniques vary so local standards should always be consulted

Source: Clean Energy Council (Australia)

disconnection switches should be installed in close proximity to the inverter. The inverter can be installed at any point in the system installation process; it may be installed at the same time as the mounting system if desired.

Installation checklist

Installers should ensure that all tools and relevant equipment are ready and available for the installation. Below is an example of a checklist for PV installations.

Table 10.2 Example of an installation checklist

Item no	Type of item	Quantity required	Details	OK
1	PV module model XYZ-90			
2	PV mounting structure			
3	Hardware for connecting module to mounting structure			
4	Hardware for connecting mounting structure to roof			
5	Hardware for ensuring roof is watertight			
6	Cable between module & PV combiner box (if one exists)			
7	Conduit if required			
8	Fastening hardware for this cable/ conduit			
9	PV combiner box (if one exists)			
10	Hardware for fastening PV combiner box to wall			

Table 10.2 Example of an installation checklist (Cont'd)

Item no	Type of item	Quantity required	Details	OK
11	Main DC array disconnect switch between solar array and inverter			
12	Cable from disconnect to inverter			
13	Conduit if required			
14	Fastening hardware for this cable/ conduit			
15	Inverter model			
16	Fastening hardware for inverter			
17	Cable between inverter and switchboard			
18	Conduit if required			
19	Fastening hardware for this cable/ conduit			
20	PV AC main switch			
21	Required signage			
22	Installation Tools (recommend technician prepares a list)			

Interconnection with the utility grid

The system in which small-scale power generators (such as rooftop PV systems or small wind turbines) are connected to the grid is referred to as interactive distributed generation.

The consumer uses electricity from both sources (PV system and utility grid) as required (as opposed to a stand-alone system where they can only use electricity produced by the PV array). This varies slightly depending on the metering arrangement (see Chapter 5 for details): when net metering is used, electricity produced by the PV system is used at that point of attachment, any excess is exported back to the grid and additional electricity is purchased from the grid when the PV system is not producing enough. If gross metering is used, all electricity is exported to the grid and the electricity required by the load is imported from the grid so that no electricity flows directly from the PV system to the load.

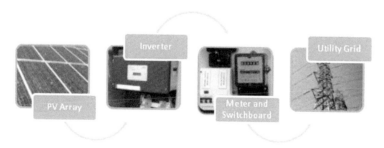

Figure 10.8 In grid-interactive systems, the PV array is connected to the inverter, which is then connected to the switchboard and then the grid; national codes and regulations cover the specifics and should always be followed

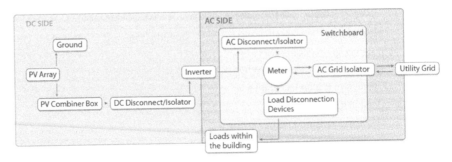

Figure 10.9 Summary of the connection of components of a grid-connected PV system using net metering

Source: Global Sustainable Energy Solutions

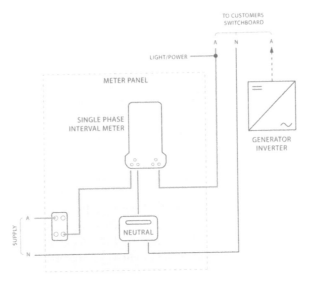

Figure 10.10 Wiring diagram showing a net-metering arrangement; only the power not used on the property is exported

Source: Global Sustainable Energy Solutions

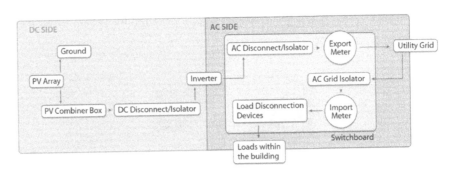

Figure 10.11 Summary of the connection of components of a grid-connect PV system using gross metering

Source: Global Sustainable Energy Solutions

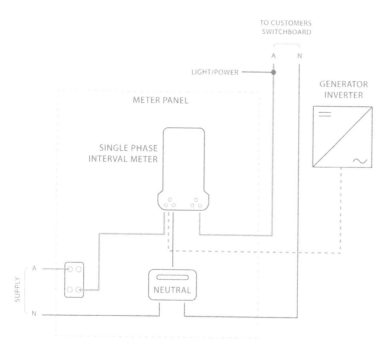

Figure 10.12 Wiring diagram showing a gross-metering arrangement where all power produced by the PV system is exported to the grid

Source: Global Sustainable Energy Solutions

In order to connect a PV system to the utility grid an interconnection agreement contract is typically required. Given that utility grid systems vary according to country of installation, so do interconnection agreements and policies. Often the agreement depends not only on local standards and regulations but also on the utility, which must agree to import the electricity produced by the PV system. Some local laws require utilities to buy the power produced by PV systems (as in the UK). However, elsewhere it is left to utilities to decide. Electricity providers worldwide have different regulations concerning safety issues that arise from connecting the grid to multiple power generators. Some of these safety issues are current overload and islanding (see Chapter 5), when electricity is fed to the grid during a power outage, which could represent a serious hazard for electricians working with power lines. Inverters are now built to prevent islanding and they should turn themselves off when the grid is down. It is not uncommon for the local utility to insist on its own inspection of the system before its initial activation.

Required information for installation

In order to install a PV system, drawings and diagrams are usually required by the local authority that authorizes the PV system installation. These drawings illustrate the on-site location of the equipment components, as well as the electrical configuration of the system. Some examples of the drawings that should be completed for each system follow:

• Electrical diagram: A simplified diagram showing the PV array configuration, wiring system, over-current protection, inverter, disconnects, required signs and AC connection to building. The wiring diagram should have sufficient detail to identify the electrical components, the wire types and sizes, number of conductors and conduit type (if needed). It should also include electrical information about PV modules and inverter(s). In addition, it should include information about utility disconnecting means (required by many utilities).

- Site plan: An architectural diagram showing the location of major components on the property. Major components of the PV system could include the array, inverter, isolation/disconnect switches, point of connection to the utility service panel. It is good practice to include major buildings/ structures on the installation site, as well as property boundaries. This drawing need not be exactly to scale, but it should represent the relative location of components on the installation site (see Chapter 7).

- Calculation sheet: Includes relevant calculations and notes related to the PV array design such as temperature-corrected maximum power and open-circuit voltage, maximum rated power, maximum power and short-circuit current. It should also list information related to the inverter such as voltage, current and power ratings, as well as calculations related to over-current protection devices.

In addition to these drawings, a permit application package should also include data sheets and installation manuals (if available) for all manufactured components including, but not limited to, PV modules, inverter(s), PV combiner box, isolation/disconnects and mounting system. Local codes and national standard will specify the documentation required and should be followed.

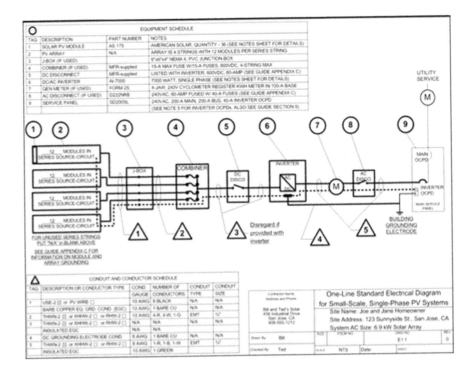

Figure 10.13 Example of an electrical diagram

Source: Brooks Engineering, www. solarabcs.org/permitting

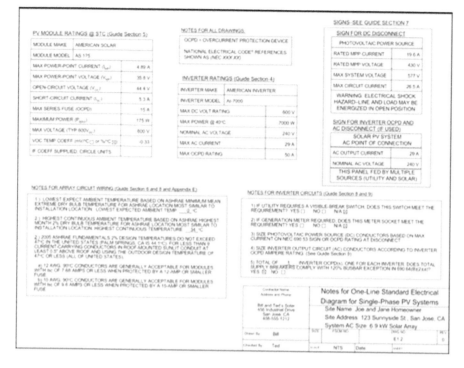

Figure 10.14 Typical calculation sheet for a PV system

Source: Brooks Engineering, www.solarabcs.org/permitting

Safety

Installing PV systems is a risky activity, so appropriate safety measures are crucial. The main risks associated with installing PV systems are:

- Electric shock: Given that PV devices generate electricity as long as light falls on them, they should always be considered electrically live and treated with due caution. In many PV systems, the array is in excess of 120V DC; voltages above this are considered deadly, for this reason local codes and national standards may require the array strings to be isolated into sections, where each section has a V_{oc} no greater than 120V DC. This is extra low voltage segmentation and is discussed in Chapter 8. Even voltages below 120V are considered very dangerous, because even if the shock does not kill the victim it may cause them to lose concentration or balance and fall off the roof if insufficiently restrained. It is very important that steps are taken to mitigate the risks of working at heights and such steps are often outlined in national safety standards or regulations. Protective clothing including gloves can be worn to protect the PV installer in case the PV array and framing is subject to an electrical fault. Potential electrical faults should be categorized and included as part of the risk assessment carried out for the specific installation.
- Working at heights: There is a high risk of falling off the roof, although if proper safety measures are in place this is greatly minimized. Each region will have its own standards for working at heights. Normally standards

require the use of safety harnesses or scaffolding when working at heights. Ladders are often used to access roofs and they should be secured. Consideration should also be given to what type of ladder is used, local codes and guidelines should be consulted as they may advise against using certain materials such as metal ladders due to their conductivity.

- Manual handling injuries: These commonly occur when people lift and carry objects that are either too heavy or using incorrect technique. It is important to bend at the knees rather than the waist when lifting or setting down an object; group lifting should also be used when required.
- Falling objects: Injuries from objects falling off the roof are also common; these can be minimized by keeping the workspace on the roof tidy and keeping tools attached to the installer at all times, i.e. by using a tool belt. If working at heights, installers may also consider restricting access to the area where they are working. For a small home installation this is generally easy because the installation only takes a few hours; however, for a large installation that will take several days this may not be possible. It is also good practice to keep any work site clean and tidy to avoid injury.
- Environment: Working outdoors for long hours (particularly during summer) can be physically exhausting. Installers should make a conscious effort to stay hydrated by drinking plenty of water and resting in the shade when needed. Sunscreen, hats and long-sleeved shirts are important to avoid sunburn. Gloves should also be worn when handling modules that have a tendency to heat up in the sun.

Before installing a PV system a risk assessment should be undertaken as part of the site assessment: risks need to identified and classified as low, medium, high or extreme depending on likelihood and severity. Risk control measures need to be identified and implemented. Advice on minimizing risks and avoiding injuries is available from the local government authority, i.e. Health and Safety Executive in the UK (www.hse.gov.uk), the Occupational Safety and Health Administration in the US (www.osha.gov), the European Agency for Safety and Health at Work (www.osha.europa.eu) and Workcover in each Australian state.

11

System Commissioning

The process of testing a PV system to confirm that it is producing electricity and interacting correctly with the electricity grid is known as system commissioning. Before an installer leaves the system to the customer it should be tested and inspected to ensure that the system is compliant with national and local standards and regulations, that all components have been safely installed and that all components are functioning as expected. Many utilities have rules or procedures that must be followed during the system commissioning process and in some cases the utility may wish to conduct a commissioning inspection. These requirements should already have been discussed with the utility when the interconnection agreement is made (see Chapter 10).

The procedures for inspecting, testing and commissioning, as well as the training requirements for the personnel who undertake these procedures, are governed by local codes and regulations. An overview of these procedures is provided in the following pages.

Box 11.1 National standards for commissioning PV systems

Commissioning procedures vary widely throughout the world and are largely dictated by national codes and standards.

- US: Article 690 of the National Electric Code (NEC) outlines inspection and testing requirements for PV systems. The Institute for Electrical and Electronics Engineers (IEEE) standard is also used: IEEE 1547 Standard for interconnecting distributed resources with electric power systems.
- International Electrotechnical Commission: The IEC provides international standards, IEC 62446 Grid-connected photovoltaic systems – Minimum requirements for system documentation, commissioning tests and inspection. This standard is intended to provide a template for electricians.
- UK: BS 7671 is the British wiring regulation and PV systems including inspection and testing requirements are covered in Section 712. The local distribution network operator (DNO) may also require various tests and documentation.
- Australia and New Zealand: The Australia and New Zealand Standards (AS/NZS) cover the commissioning of photovoltaic systems in AS/NZS4777

Grid connection of energy systems via inverters and AS/NZS5033 Installation of photovoltaic (PV) arrays.
- Canada: The Canada Standards Association covers commissioning in CSA C22.2 no. 107.1-01 (R2006) – General use power supplies.
- Singapore: CP 5: 1998 – Electrical installations and Amendment no. 1 to CP 5: 1998

Final inspection of system installation

Before the PV system is commissioned, a final inspection should be undertaken to ensure the system is ready to be tested. If any issues are identified they should be addressed before any part of the system is switched on and/or tested. The equipment and installation should be checked to ensure that:

- Equipment and components are not damaged.
- The system matches the design documents and all equipment has been correctly connected according to the wiring diagrams.
- Equipment and components comply with local safety standards and are suitable for use in a utility-interactive PV system.
- The site has been left clean and tidy and presents no hazards for the general public.
- The signs and warning labels required by local codes are present.

The array and array frame should be inspected to ensure that they have been installed correctly and are suitable for the location. This includes checking that the frame is sturdy, is appropriately rated for local wind and snow conditions, and has been installed so that any roof penetrations are properly sealed and weatherproofed (see Chapter 6).

The inverter should be inspected prior to commissioning to ensure that it is securely mounted and all electrical connections in and out of the unit are firm. The location of the inverter should also be considered during this inspection to ensure that it is accessible for maintenance and emergency disconnection, has been appropriately weatherproofed, and that allowances have been made to ensure sufficient ventilation. Weatherproofing can be verified by checking the IP or NEMA rating of the inverter (see Chapter 8 for further information about this). Ventilation should also be checked against the manufacturer's recommendations.

The wiring and electrical components should also be inspected. The inspection should ensure that all wiring and components are securely installed and adequately protected against mechanical and environmental damage. It should be ensured that they are fully operational, are correctly sized and installed in accordance with standards and regulations. All disconnects/isolators must be easily accessible in case of an emergency.

This inspection process should be documented and a copy of the documentation should be left with the customer for their records.

Box 11.2 System installation and pre-commissioning checklist

PV array

Mounted flat on roof
Building integrated ☐

☐
Mounted on tilted array frame ☐
PV array tilt ……………………………………..
PV array orientation ………………………..
Solar array is securely fixed ☐ Details…………………………………………………........
No dissimilar metals are in contact with the array frames or supports ☐
Roof penetrations are suitably sealed and weatherproofed ☐
Wiring is protected from UV and mechanical damage ☐

Inverter

If required by local code, isolating device is mounted close to the inverter ☐
If required by local code, isolating device is mounted on output of inverter ☐
Inverter is housed in weatherproof enclosure or inside building ☐
Adequate space and ventilation for inverter ☐

Low voltage DC cabling

Is clearly identified in accordance local guidelines ☐

Signage (white on red)

Note: The following examples are from Australian and New Zealand Standards (AS/NZS 4777.1: 2005 Grid connection of energy systems via inverters – Installation requirements and AS/NZS 5033: 2005 Installation of Photovoltaic (PV) arrays)

The following sign is permanently fixed on the switchboard ☐

WARNING

DUAL SUPPLY

ISOLATE BOTH NORMAL AND SOLAR SUPPLIES BEFORE
WORKING ON THIS SWITCHBOARD

The following sign is permanently fixed to main disconnect/isolator   ☐

| Normal Supply |
| MAIN SWITCH |

The following sign is fixed adjacent to the PV array main disconnect/isolator in switchboard   ☐

| PV ARRAY |
| MAIN SWITCH |

If the solar system is connected to a distribution board, then a system-specific fire emergency information sign (such as the following) is located on main switchboard and all intermediate distribution boards   ☐

| SOLAR ARRAY ON ROOF |
| Open-circuit voltage: 220V |
| Short-circuit current: 20A |

If the solar system is connected to a distribution board, then following sign is located on main switchboard and all intermediate distribution boards   ☐

WARNING

DUAL SUPPLY

ISOLATE SOLAR SUPPLY AT DISTRIBUTION BOARD

The following is permanently fixed on PV combiner boxes   ☐

| SOLAR DC |

A detailed, system-specific, shutdown procedure (such as the following) is permanently fixed on the main switchboard   ☐

SHUTDOWN PROCEDURE
1. Turn off the inverter AC Main Switch located next to the output terminal of the inverter.

2. Turn off the PV Array Main switch located next to the input terminals of the inverter.

WARNING: Do not open plug and socket connectors or PV string isolators under load.

PV Array open-circuit voltage: 220V

PV Array short-circuit current: 20A

Authorization: I, ...
verify that the following system has been installed to the standard indicated by
these guidelines and complies with the relevant standards.
Name of the person for whom the system was installed

....

Location of system ....

....

Signed ....
Date ....

Testing

Following a visual inspection of the system, testing should be undertaken in accordance with the prevailing national codes. National codes may require installers to ensure that the following points are compliant prior to system testing:

- There is no voltage at the output of the PV array (and at the output of each string if there is more than one). This may be achieved by leaving one of the module's interconnects disconnected/unplugged.
- Any fuses have been removed and all circuit breakers are in the 'off' position.
- The AC and DC main disconnects/isolators are in the 'off' position; local codes may also require them to be tagged or locked for the duration of the testing procedure.
- All components, i.e. the inverter, are switched off.

After the safety requirements outlined in the local codes have been fulfilled, testing may be undertaken. Each component is switched on and each isolator closed individually, beginning at the array and ending at the loads (i.e. appliances). Testing is done in this order for safety; it reduces the risk of hazards and equipment damage should any problems occur. At each step the system is tested by measuring the system parameters using meters and the displays on various components (i.e. the inverter display will show important data about the system). If at any stage the system begins to operate outside the expected parameters then electricians must identify the problem and address it before testing can continue. Common parameters that are tested include:

- continuity between adjacent system components;
- resistance of cable insulation;
- measurement of array and string open-circuit voltage (a large difference in the open-circuit voltage of identical strings or an open-circuit voltage very different to that expected may indicate a problem);
- measurement of array and string short-circuit current (hazardous – see box below);

Figure 11.1 An electrician checks the open-circuit voltage and polarity of an array cable entering the inverter

Source: Global Sustainable Energy Solutions

- measurement of voltage drop across string fuses;
- verification of polarity of installed components;
- grounding/earthing of the system, often including impedance loop test or ground electrode test for which a ground tester is required.

Box 11.3 Measuring short-circuit currents of strings and array

WARNING: The following procedures, which are extremely hazardous if not carried out correctly, should only be conducted by suitably qualified personnel. The procedures describe how to measure short-circuit currents. Voltages can be very high and if the procedures are not followed, then arcing and damage to components could occur.

Note: In some projects, short-circuit currents are required to be recorded as part of the contractual commissioning. If it is not necessary to know the short-circuit current for the system, then recording the operating current of each string is sufficient. This can be done by using the meter on the inverter or by using a clamp meter when the system is in operation.

A: Where short-circuit currents are required

1   Ensure each string fuse (if installed) is not connected or that LV arrays are still broken into ELV segments.
2   Leave the solar array cable connected to the PV array DC isolator.
3   Remove the cable from the PV array DC isolator to the inverter.
4   With the DC isolator off, put a link or small cable between the positive and negative outputs of the PV array DC isolator.
5   Install the string fuse or reconnect the ELV segments for String 1. Turn on the PV array DC isolator. Use a DC clamp meter to measure the DC short-circuit current for String 1.

6   Turn off PV array DC isolator. Disconnect string fuse for String 1.
7   Repeat step 5 for each individual string.
8   After the DC short-circuit current for each string has been individually measured, check that the PV array DC isolator is off, and then install all string fuses or connect the ELV segments in each string. Turn on the DC switch and measure the DC array current using a clamp meter. Turn off switch and remove link in output of PV array DC isolator.

B: Where short-circuit currents are not required

1   Wait until the system is connected and the inverter is operating. If there is only one string in the array, record the operating current of that string.
2   If there is more than one string, turn off the inverter, the inverter AC isolator and PV array DC isolator. Isolate all the strings by either removing fuses, turning off the circuit breaker or using a disconnection device.
3   With one string connected at a time turn the system back on and record the operating current of that string.

Note: These tests should be performed on a bright, sunny day with no cloud. This is to avoid variable readings due to cloud cover. Ensure that all occupational health and safety precautions are taken when carrying out these tests; particularly for working at height, in bright sunlight, on a hot surface in hot and sunny conditions.

Box 11.4 Pre-commissioning test sheet

PV array

There is no voltage on input side of PV combiner box   ☐
There is no voltage on output side of PV combiner box   ☐

Continuity between strings and the PV combiner box

String 1 +ve   ☐
String 1 −ve   ☐
String 2 +ve   ☐
String 2 −ve   ☐
String 3 +ve   ☐
String 3 −ve   ☐

Correct polarity between strings and PV combiner box

String 1   ☐
String 2   ☐
String 3   ☐

WARNING: If polarity of one string is reversed it may cause a fire in the PV combiner box.

Open-circuit voltages

String 1   V
String 2   V
String 3   V

Continuity between PV combiner box and DC main disconnect/isolator

Array +ve   ☐
Array −ve   ☐

Correct polarity between PV combiner box and DC main disconnect/isolator   ☐

Short-circuit currents (if required)

String 1   A
String 2   A
String 3   A
Array   A

Continuity between DC main disconnect/isolator and inverter

Open-circuit voltage at input of DC main disconnect/isolator   V

Array +ve   ☐
Array −ve   ☐

Correct polarity between PV array DC isolator and inverter   ☐

Inverter

Continuity between inverter and kWh meter

Line   ☐
Neutral   ☐
Correct polarity between inverter and kWh meter   ☐

Continuity between kWh meter and AC main disconnect/isolator

Line   ☐
Neutral   ☐

Correct polarity between kWh meter and AC main disconnect/isolator   ☐
Correct polarity at output of AC main disconnect/isolator   ☐

Voltage at output of AC main disconnect/isolator   V

Initial reading of kWh meter   ...

Commissioning

Commissioning will be the first time the complete system is switched on and able to feed electricity into the grid. As already mentioned this process is normally covered in great detail by national codes and standards, with which any electrician undertaking PV system commissioning should be very familiar. The following table outlines the necessary post-commissioning tests to ensure the system is functioning with the grid and operating as expected. It is particularly important that the PV system disconnects and reconnects to the grid in accordance with local standards. This is required to ensure that an islanding situation does not occur (see Chapter 5).

Box 11.5 Commissioning test sheet

Refer to system manual for the inverter and follow start-up procedure
System connects to grid   ☐

When inverter, AC main disconnect/isolator and DC main disconnect/isolator have been turned on and inverter start-up procedure followed:

Voltage at DC input of inverter   V

Voltage within operating limits of inverter   ☐

Voltage at AC output of inverter   V

Input power of the inverter (if available)   W

Output power of the inverter (if available)   W

Output power as expected   ☐

System disconnects from grid when inverter AC isolator turned off   ☐

System documentation

At the completion of the installation the owner should be supplied with a system manual that includes information on the system. Local codes generally specify the documentation that should be provided; however, a general guide is given below.

- List of equipment supplied: the system manual should include a full itemized list of all the components that have been installed including PV modules, inverters, array frames, PV combiner boxes, string isolators, fuses or circuit breakers and the DC and AC main disconnects/isolators. The list should include the quantities of equipment items used.
- System diagrams: a basic circuit diagram and a wiring diagram should be included in the manual. Architectural drawings or the site plan showing the major components are also useful.
- System performance estimate: the manual should include the expected yield of the system as calculated by the designer. It may also include information on local financial incentives (see Chapter 13) and what these mean for the system in terms of income and/or savings on electricity bills. It is also important to emphasize that this is purely an estimate and some variation from year to year is common.
- Operating instructions for the system and its components: the manual should include a brief overview of the system, the function of each of the main components and how the system operates. Any information important to that particular system should be included in the manual. It is important to explain to the owner that the system will turn off when the grid fails, i.e. when there is no power available from the grid.
- Shutdown and isolation procedures for emergency and maintenance: the manual should include procedures for maintenance and emergency. Depending on the size of the system, maintenance procedures might not involve the complete shutdown of the system.
- Maintenance procedure and timetable: Chapter 12 details the maintenance requirements for a grid-connected PV system, including tables that should be included in maintenance logbooks. This information should be incorporated into the system manual.
- Installation and commissioning records: these should be all signed and included in the system manual.
- Monitoring of system: a section of the manual should advise the system owner how to monitor the system to ensure that it is operating correctly. Many inverters have monitoring capabilities. If the inverter includes these features, instructions on how to use them should be provided. If a separate monitoring unit has been supplied with the system, the manual must include information on its operation.
- Warranty information: a grid-connected PV system comprises individual products connected together by a system installer to create the system. There are four types of warranties applicable to such a system, these are:
  1 product warranties covering defects in manufacture;
  2 product warranties related to output performance over time;

3   system warranties relating to proper operation of the installed system over time;

4   energy performance warranties relating to the guaranteed energy output of the grid-connected PV system over a period of time, typically one year (see Chapter 3).

The first two warranties in this list are the responsibility of the equipment manufacturer, but a system owner could contact the installer for help if a warranty claim is required. The last two warranties are provided by the system installation company. Details of all the warranties offered should be included in the system manual.

• Equipment manufacturers' documentation and handbooks: all product manuals provided by the various manufacturers should be included in the system manual. Examples include inverter manuals, PV module data sheets and technical information on any balance of system (BOS) equipment.

12

PV System Operation and Maintenance

System maintenance

This chapter outlines the maintenance required for a grid-connected PV system. This information is of a general nature, and it is important to follow the system owner's manual and the manufacturer's recommendations for maintenance of the equipment. Some simple maintenance tasks may be undertaken by the owner, but most should be performed by a qualified PV technician. Many regions require testing of electrical installations (such as a PV system) every few years and the installer should be familiar with local regulations.

A maintenance contract is often included in an installation package: the company will return once or twice a year to inspect and test the system in order to ensure it is functioning properly. Results of these inspections are recorded in a maintenance schedule and equipment logbook that should have been provided to the system owner before commissioning the system. In addition the maintenance contractors should keep a record of service and repair work, a copy of which must also be retained by the system owner. Good record keeping is essential in the PV industry because the products used are covered by very long warranties. Logbooks can be particularly useful because this historical information can show changes over time, as well as variations from the norm, indicating a problem or one in the making.

The most important aspect of system maintenance is performance monitoring, which will ensure that when a problem arises it can be identified and corrected before serious damage occurs (shading causing hot spot heating, for example). It will also give the user records of year-round system performance, which will help performance comparison in the future.

PV array maintenance

PV arrays have no moving parts and therefore operate for many years, often without issues. However, regular maintenance is still important to ensure a PV array is operating safely and efficiently. Common maintenance tasks include cleaning the modules and mechanical and electrical checks. It is recommended that PV modules are maintained only by suitably trained service personnel.

PV modules are generally self-cleaning; rain washes away any dirt, but a build-up of dirt, bird droppings or leaves needs to be dealt with. Solvent should never be used to clean a PV array; fresh water with a small amount of soap is

Figure 12.1 Dusty modules will have a reduced yield

Source: Global Sustainable Energy Solutions

sufficient. Snow on arrays is usually not a problem as long as the modules comply with the relevant standards (see Chapter 3); however, if much snow is present for long periods of time the owner may wish to consider clearing it as it will shade the array and affect performance. System owners should be aware of and account for the dangers of working at heights before attempting to clean the array themselves.

It is also important to check that the array is not being shaded by any trees that may have grown since the system was installed; frequent pruning may be required. Trees on neighbouring properties may also shade the array. Some

Figure 12.2 Cleaning a PV module involves spraying them with a hose and occasionally scrubbing to remove baked-on marks

Source: Brian Kusler

regions have laws to protect solar access, but many do not and it is important to discuss this issue with neighbours before the system is even installed.

The maintenance undertaken by a qualified PV technician will normally include an inspection of the mounting structure for signs of corrosion that may lead to mechanical failure and the points of attachment to the roof should be inspected to ensure they are still watertight. The cabling and wiring should also be inspected to check it is mechanically sound and weatherproof.

It is also common practice to monitor PV systems. Although it is normal for power output to degrade over time, a large difference between measured and expected values indicates a problem to be tackled. The most common form of monitoring is checking electrical parameters such as output voltage and current, which can be compared with what is expected from the design and what was measured during the commissioning process, to determine whether there is a problem or not. Many modern inverters include monitoring functions, or external monitoring equipment may be used (see Chapter 5). In some cases the environment may also be measured (temperature), as this helps designers see a correlation between environmental conditions and system yields. The temperature of the modules may be monitored to ensure it is within the designed system range (see Chapter 9).

In order to be interpreted, monitoring data must be somehow displayed. Basic monitoring systems may involve a small LED display and indicator lights on the component, to be checked against that component's user manual. Most modern inverters have small display screens with measured values and many have wireless connections to a remote display (such as a computer or iPhone). Monitoring information can then be recorded to show trends over time, which is often of great interest to the customer who likes to know how much the system is producing and how much they are saving (or making if a feed-in tariff is present).

An array maintenance sheet should be included in the logbook.

Table 12.1 Recommended maintenance for PV array

Activity	Frequency
Clean modules	As required
Check mechanical security of the array structure	Annually
Check all cabling for mechanical damage	Annually
Check output voltage and current of each string of the array and compare to the expected output under existing conditions	Annually
Check electrical wiring for loose connections	Annually
Check the operation of the PV DC main disconnect/isolator (only after AC main disconnect/isolator has been switched off)	Annually

Table 12.2 Example of a PV array logsheet

Date	Cleaned modules	Array structure OK	Array cabling: mechanical	Array cabling: electrical	Output voltage	Output current	PV DC main disconnect/ isolator	Comments
	☐	☐	☐	☐	...V	...A	☐	
	☐	☐	☐	☐	...V	...A	☐	
	☐	☐	☐	☐	...V	...A	☐	

Box 12.1 Example of maintenance instructions from Suntech Power

Suntech recommends the following maintenance in order to ensure optimum performance of the module:

- Clean the glass surface of the module as necessary. Always use water and a soft sponge or cloth for cleaning. A mild, non-abrasive cleaning agent can be used to remove stubborn dirt.
- Check the electrical and mechanical connections every six months to verify that they are clean, secure and undamaged.
- If any problem arises, have it investigated by a competent specialist.
- Observe the maintenance instructions for all components used in the system, such as support frames, charging regulators, inverters, batteries etc.

Module manufacturers will also have their own maintenance instructions, and these should be the first thing consulted as non-compliance may void warranties.

Inverter maintenance

Inverter maintenance generally involves:

- keeping the unit clean: the housing and any air vents should be cleaned when required;
- ensuring the unit is not invaded by insects and spiders;
- ensuring all electrical connections are clean and tight;
- checking the operation of the inverter.

Inverter maintenance operations and basic repair instructions are likely to be listed in the manual. If any maintenance is undertaken on the inverter, it should be recorded in the system logbook. Inverter maintenance (aside from cleaning) and repairs should only be carried out by a suitably qualified technician; system owners are not to attempt this.

System integrity

The above maintenance checks relate to the individual components of a system. For them to work as a system they need to be interconnected by cables. It is therefore essential during any equipment maintenance that the whole system should be visually checked to ensure that there is no potential threat to its performance and/or safe operation. This involves inspecting and testing (where appropriate) every component including isolators, circuit breakers, PV combiner boxes and wiring.

Troubleshooting

A well-designed and well-installed PV grid-connected system should have fault-free operation for many years, but problems can occur. If the system is not working, it is usually due to one of two reasons: either a piece of equipment is faulty or there may be a grid problem. The cause of the problem must be discovered and fixed. Troubleshooting should only be undertaken by a suitably qualified PV technician as it can involve risks such as working at height and high voltage. This section discusses fault-finding and problem-solving in some common situations.

Identifying the problem

By monitoring system performance, a system owner will notice that the system is not working or underperforming. If the problem is not immediately obvious from a visual inspection, i.e. shading of the array, then the owner should contact the installer, who should visit the site.

The troubleshooting visit should occur during daylight hours, preferably with enough sunshine to be able to test the output of modules and with sufficient daylight hours to allow a full investigation. The first thing to check is that neither the array DC main disconnect/isolator nor the array AC main disconnect/isolator has been switched off.

If all the disconnects/isolators are on, then a visual check of the inverter should be undertaken. If it has failed, it is usually caused by failure in the electronics. This might be accompanied by an error message or other displayed message. If the inverter has failed then a decision will need to be made about whether it can be repaired at the site or returned to the manufacturer. If the inverter is not on, but the problem is not immediately obvious, i.e. there are no error messages, it is normally necessary to determine whether there is AC power at the inverter and then whether there is DC power at the inverter. If there is no AC power from the grid, then the system must be systematically checked all the way back to the point of supply to find the fault. This will involve measuring whether AC voltage is present:

1 at the meters; then
2 on the inverter side;
3 followed by the supply side of the array AC main disconnect/isolator; then
4 at the point of attachment to the grid supply.

If there appears to be no fault at the inverter and the AC power is connected to the inverter, then the array should be investigated.

Troubleshooting PV arrays

This process varies depending on the configuration of the system. The process described here assumes there is a PV combiner box between the array and the DC main disconnect/isolator. If the fault appears to be on the DC side of the inverter (that is on the PV array side) the probable cause is that either the inverter is not receiving sufficient power from the array and so has switched off or that the system is underperforming because the array is not producing the expected energy. Any electrical service or maintenance work carried out on PV arrays must be done by suitably qualified personnel.

If it appears that there is a fault on the DC side of the inverter, the first task is to measure the PV array open-circuit voltage at the input terminals of the inverter and confirm that DC power is reaching the inverter. If there is no DC power at the inverter, then the technician should systematically check each component and connection (by measuring if DC voltage is present) in order, starting at the inverter and ending at the PV array.

If there is no DC voltage at the inverter it is common practice first to check whether the DC main disconnect/isolator is operating and that the circuit breakers or fuses (if present) are closed and operational. If there is no DC voltage at the PV combiner box, then the fault is within the array. This generally indicates that one (or all) the array strings have failed. Possible causes of string malfunction include:

- cable disconnection or plug failure (if plugs are used);
- a loose connection within the PV combiner box;
- a failed or broken module.

The actual fault may only be found by physically checking the cables and individual modules and taking current and voltage measurements. Zero current indicates a fault.

Troubleshooting underperforming systems

Troubleshooting is also required when the system is operating but underperforming. If the system is underperforming, the following steps should be undertaken:

- visual inspection of the system to check for shading or other immediately obvious faults such as a damaged module;
- measurement of the current from the array using a clamp meter.

If there is no shading problem, but the current is lower than expected, the array should be tested by taking measurements. If there are multiple strings, each one should be turned off systematically. If one of the strings is not providing power, then there will be no change in the current when that string is turned off. Once the faulty string has been identified, the circuit breakers or fuses should be checked for failure. If they have not failed, the steps outlined in the previous section should be followed.

Box 12.2 Undertaking fault-finding on the PV array

It is important to have a basic understanding of how the module performs:

- The open-circuit voltage and short-circuit current can be measured only when there is no load on the modules.
- The modules produce close to open-circuit voltage even when the solar irradiance is low and the short-circuit current is proportional to the available solar irradiance.
- A shaded module will produce either no current or reduced current depending on the shading.
- If one module in a long string is shaded, the operating current of the string might not change due to the bypass diodes operating, but the open-circuit voltage will be reduced. See Chapter 4 for operation of bypass diodes.

Troubleshooting inverters

If there are DC voltage and AC voltage at the inverter terminal (and the PV array output is sufficient) yet the inverter is not operating, then the inverter has possibly failed. Most manufacturers include in the manual a troubleshooting section that should be consulted. In many cases the inverter will display fault lights or error messages to indicate the problem; the manual should be used to identify and address these errors. Some common problems are:

- Grid voltage is too high or too low: this indicates a fault with the grid and so the electricity distributor should be contacted.
- Grid frequency is out of range: this indicates a fault with the grid and so the electricity distributor should be contacted.
- DC voltage from the array is too low: the array may require troubleshooting.
- DC voltage is too high: the array should be immediately disconnected as it may damage the inverter.
- Line impedance is too high: the connections on the AC side should be checked as they may be loose. If the problem persists, it may have something to do with the grid and so the electricity distributor should be contacted.
- Earth leakage current is too high (for transformer-less inverters only).

Other common problems

Other common problems with PV arrays outside these technical categories include, first, a client's unrealistic expectation of their system's performance. Unfortunately little can be done to rectify this, so it is important to have good communication with the client so they understand the system capabilities before installation. This problem can also occur when the system designer has made an error and not appropriately accounted for all the system losses; hence overestimated the system's output (see Chapter 9). Such calculations should always be undertaken conservatively and the reasons for system underperformance should be clearly explained to the client.

Poor system design is also a major cause of PV system failure. This often happens when the PV array has not been correctly sized for the inverter used (see Chapter 9) and so frequently the PV array produces voltages or currents outside its input windows, forcing the inverter to shut down or disconnect from the grid. In these cases the fault in the system design must be corrected, which may involve the purchase of replacement components and the services of an installation team. If the voltage falls below the inverter's voltage window it is likely that high temperatures have caused a reduction in voltage which is larger than expected (see Chapter 4). If the array is not well-ventilated this can be rectified by modifying the mounting structure (see Chapter 6) to allow for convective cooling behind the modules. This will lower the modules' operating temperature and decrease the reduction in voltage that occurs at high temperatures. The other option is to add more PV modules in series to each string; care must be taken when doing this to ensure that the system voltage will not then exceed the maximum inverter input voltage on very cold days (see Chapter 9). If the fault is caused by over-voltage then it may be necessary to remove modules from the string or strings in order to reduce overall system voltage. Over-voltage is a serious fault that can damage inverters. The inverter manufacturer will specify the voltage operating window for their inverter and some inverter data sheets will also show a 'maximum DC voltage' figure that is higher than the maximum operating voltage figure. The inverter and PV module array specifications must be confirmed ahead of time to avoid this.

Unstable grids may also cause system failure. These problems usually only last for a short time and are difficult to identify without the help of the utility. Fluctuations in grid voltage and/or frequency may mean the grid operates outside the inverter's AC voltage and/or frequency windows, so the inverter shuts itself off. Most metropolitan grids are fairly stable, but people living in rural areas at the end of long transmission and distribution lines may experience grid failures more frequently. The electricity utility managing the local grid should be contacted in this case.

13

Marketing and Economics of Grid-connected PV Systems

Solar radiation is free, but it can be expensive to set up a PV system capable of taking advantage of this free energy source. Conversely, energy from fossil fuels is not free (i.e. you need to continually buy fuel) but the systems using these energy sources are already well established, so at this point in time it is less expensive to continue using fossil fuels than to build new infrastructure to integrate renewable energy generators into the grid.

Despite this, people continue to invest in grid-connected renewable energy for a variety of reasons and one of these is economics. This chapter discusses the economics of PV systems and demonstrates how to calculate payback time (i.e. the number of years the system will take to pay itself off). Like any emerging technology, PV requires support in getting off the ground; with every system installed the industry is learning and creating a competitive and innovative market that will drive prices down. The support currently available to PV is often in the form of financial incentives available from governments and utilities. These greatly decrease the payback time and can make owning a PV system a profitable and secure investment. Common types of financial incentive available are discussed in this chapter. Marketing PV systems is also covered, including the unique advantages of PV systems, common barriers to the widespread use of PV and ways in which these are being overcome.

Ultimately the PV industry is aiming for grid parity; grid parity occurs when the cost of electricity from a PV system is equal to the cost of electricity from the grid to the final consumer. Grid parity is highly dependent on local factors such as the cost of PV systems and installation, local solar radiation and local electricity prices, and so will occur at different times in different countries; however, it is expected to occur in many developed nations within the next decade.

As the financial benefits of installing a system are a priority for many people, it is very important that the designer is familiar with local PV system incentives and able to give the customer an accurate assessment of system payback. A list of sites containing information about the financial incentives offered for PV in major English-speaking nations is given in Chapter 15.

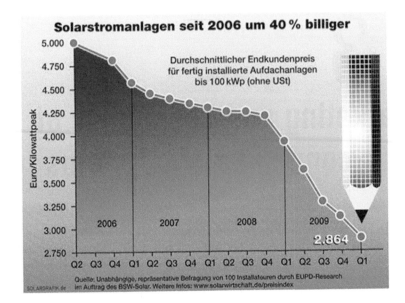

Figure 13.1 The price of PV systems is declining rapidly, making them a more attractive investment. The cost of small-scale PV systems has been reduced by 40 per cent in the last four years

Source: www.solarwirtschaft.de

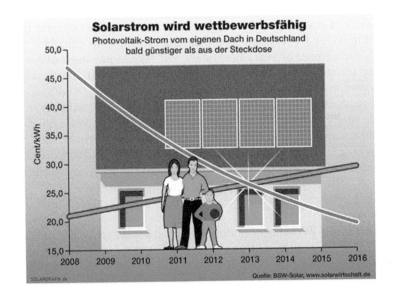

Figure 13.2 PV is expected to achieve grid parity in Germany within five years. Contributing factors include the strength and competitiveness of the local installation industry and the relatively high electricity prices

Source: www.solarwirtschaft.de

PV system costing

The costs associated with the individual components of the PV system need to be analysed in order to estimate the full cost of the system. System costs will vary significantly depending on the area and the local PV market (larger markets with many different companies operating tend to offer lower prices due to high competition). The main costs associated with a PV system are as follows:

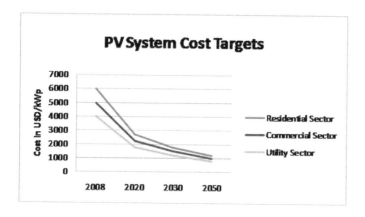

Figure 13.3 Global average PV system capital costs per kWp installed and their expected decrease over the next four decades

Source: International Energy Agency (IEA)

- Capital costs: The upfront purchase of all system equipment including PV modules and balance of system equipment makes up approximately 80 per cent of total system costs. The modules are the most expensive item by far, but inverters can still be costly; grid-interactive inverters range from US$500/kW to US$2000/kW. Smaller inverters generally used in residential applications are in the more expensive end of the spectrum because they are smaller. In areas where utilities are obliged to offer free net metering (as is the case in most US states) to customers with a PV system a new meter can be acquired free of charge, but in areas where the policy does not exist (i.e. Australia) it is often necessary to purchase a new meter with the system. The remaining 20 per cent of system cost is for the actual installation (excluding a small ongoing maintenance cost).

- Maintenance costs: 1 per cent of system cost is comprised of maintenance costs. Maintenance should be undertaken every 6–12 months; if the modules and inverter have been installed correctly then maintenance costs should be minimal. The array should last at least 20 years, and most modules are covered by a 20–25 year warranty so if premature failure does occur it is often possible to replace or repair the modules free of charge. The inverter is likely to require repair during its lifetime; inverters generally carry a 5–10 year warranty with an option to extend the warranty a further 5–10 years.

- Replacement costs: PV modules are expected to last at least 25 years and most system components are expected to last at least 20 years. Some system components may not last as long as the panels and will require replacement. Inverters have warranties that commonly last 5–10 years but can mostly be repaired should they fail beyond this period. Terminal failure of a correctly installed and properly sized inverter is uncommon, but possible. The designer should ask the manufacturer about the expected life of the inverter they are installing in the system. If it is less than the expected life of a system, then a replacement should be accounted for in costing. Other components may require replacement: monitoring equipment, bypass diodes, cable, plugs/sockets etc. Protection from the elements and wildlife will increase the lifetime of this equipment and hence reduce these costs.

Valuing a PV system

Valuing a PV system is an important process that allows for comparison across systems. The preferred method for assessing the capital cost of a PV system is in dollars per watt ($/W) and as such looks only at the upfront cost of the system. To calculate $/W the following formula is used:

$$\$/W = \frac{\text{Upfront cost of the PV system (\$)}}{\text{Rated Peak Power of the PV System (W)}}$$

This method is only used for upfront costs. Currently it is the standard method of comparing system and equipment costs in Europe and is becoming increasingly common around the rest of the world.

Simple payback and financial incentives

PV systems are an investment and it is often desirable to calculate the payback time, i.e., the number of years a system will take to pay itself off. This section will outline how to calculate payback period and the financial incentives that can improve the payback time.

Simple payback

The simplest method for examining the economics of PV grid-connected systems is the *simple payback* method. People are often interested in the payback period of the installation. This is calculated using the following formula:

$$\text{Time (years)} = \frac{\text{Capital cost (\$)}}{\text{Savings from avoided electricity purchase (\$)}}$$

Example:

A 1kWp grid-connected system produces about 1200kWh per year. The system costs currently $6,000.00. The average projected costs for residential electricity is $0.15/kWh, therefore the savings per year will be $180 (1200 × 0.15).

The simple payback time is: $T = \dfrac{\$6000}{\$180} = 33 \; years$

Feed-in tariffs

A Feed-in tariff (FiT) is a monetary reward for feeding electricity generated by PV into the grid. It can either be equal to the retail electricity rate or greater than this rate (known as an enhanced FiT). FiTs are usually financed by a levy added to all electricity bills. Small-scale PV is generally most successful in locations that have FiTs, such as Germany; however, it is important for these FiTs to be stable in order to encourage sustainable growth in the industry. In

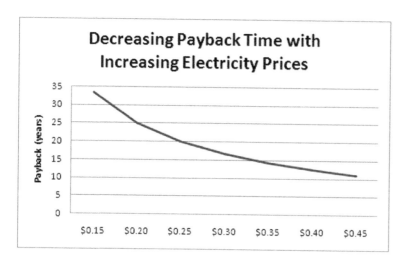

Figure 13.4 Payback times will decrease with the inevitable rise in electricity prices even if PV system prices do not decrease

both Spain and Australia there have been recent cases where an FiT was introduced by the government but removed within a year. FiTs can be structured in two different ways:

- Gross FiT: All electricity generated, regardless of whether it is used in the customer's house or not, receives the FiT.

or

- Net FiT: Only electricity that is exported to the grid receives the FiT. Some net FiTs are time-of-use FiTs, i.e. if the customer is away from home during the day while the PV system is producing electricity, then the customer receives the FiT (even if they use more energy at night than they produced during the day). Other net FiTs go by the total amount of energy produced vs. the amount consumed; i.e. when it was consumed is irrelevant.

Box 13.1 Calculating payback time for PV systems receiving an FiT

Example:
The following table contains details about the UK gross FiT known as the Clean Energy Cash Back scheme.

The payback period for a 3kWp system installed on a new building in the UK in July 2011 is calculated: (the system is expected to produce 2700kWh/year and the retail price of electricity is £0.13/kWh)

Purchase price of system: £15,000

Capacity installed	Tariff level for new installations in period (£/kWh)			Tariff lifetime
	Year 1	Year 2	Year 3	
	1 April 2010 to 31 March 2011	1 April 2011 to 31 March 2012	1 April 2012 to 31 March 2013	
Less than 4kW (new building)	£0.36	£0.36	£0.33	25
Less than 4kW (retrofit)	£0.41	£0.41	£0.38	25
4–10kW	£0.36	£0.36	£0.33	25
10–100kW	£0.31	£0.31	£0.29	25
100kW–5MW	£0.29	£0.29	£0.27	25

Savings per year: 2700kWh/year × £0.13/kWh = £351
Income from FiT: 2700kWh/year × £0.36/kWh = £972

$$\text{Time} = \frac{£15,000}{£(351 + 972)} = 11.3 \text{ years}$$

After 11 years the owner will begin to make a profit on their system and this will be guaranteed for the following 14 years (because the FiT lifetime is 25 years).

Box 13.2 Important features of FiTs

- Rate: some FiTs such as the UK FiT shown above give generators a fixed price for electricity, but other FiTs may vary as the retail price of electricity varies so they are always paying the same premium on the retail electricity price.
- Lifetime: FiTs are usually guaranteed for a certain number of years. It is important to check that the system will actually pay itself off in that time period before investing in a PV system. Lifetimes may vary from as little as 7 years to as long as 25 years.
- Conditions: there may be conditions attached to receiving the FiT, e.g. the generator may be required to surrender their Renewable Energy Credits (explained later in this chapter) to the utility purchasing their electricity.
- Programme end date: FiTs make PV systems highly profitable. But as they can also be costly schemes, most FiTs are regularly reviewed by the governing body after a certain time period or when a certain amount of PV has been installed under the scheme. It is important to make sure the scheme is still available before buying a system.

Rebates

Rebates are a financial incentive intended to reduce the upfront cost of a PV system and are usually a one-off payment. This type of incentive is usually available to smaller systems such as those on houses or urban buildings. A rebate can significantly reduce the payback time of a system, but they are often only available for a short time, so it is important to check that the rebate is still available before installing the system. This information should be available from the utility, local, state or federal government.

Rebates generally reward capacity installed, the rebate is calculated from a $/W installed price. An example is the Tennessee Solar Institute Solar Installation Grant Program; a table of the rebate prices available for PV systems is given below. The programme has a US$9 million budget and will end when funding is exhausted, it also has a maximum incentive size of US$245,000.

Box 13.3 Calculating payback time for a PV system receiving rebates

Example:

Using the first example's system, calculate the payback time for a system located in Tennessee under the state rebate programme:

Capacity installed	Rebate
1–30kW	$2.00/W
30–60kW	$1.50/W
60–200kW	$1.00/W

The system is 1kWp, so the value of the rebate would be: 1000W x $2.00/W = $2000.

Subtract this from the capital cost of the system ($6000) and the net cost is $4000.

Divide this by the money saved due to reduced electricity usage, calculated in the first example to be $180/year:

$$\text{Time} = \frac{\$4000}{\$180} = 22 \text{ years}$$

This rebate reduced the payback time of the system from 33 years to 22 years; however, that is still significantly longer than the payback time for systems in areas with FiTs.

Box 13.4 Important features of rebate schemes

When planning to claim a rebate it is important to look for several key pieces of information:

- Programme end date: all rebate programmes eventually draw to a close. A programme will generally end at either a specific predetermined date, when a certain goal of capacity installed is achieved or when the budget allocated to the programme is used up. It is important to be aware of these factors and ensure the rebate has been applied for before the programme closes.
- Size constraints: ensure the system is within the size limit of the rebate programme; many rebate programmes will also have a lower incentive for larger systems.
- Conditions: other conditions may be present, i.e. the system may need to be installed by an accredited electrician to be eligible for the rebate.

Tax incentives

In some areas (particularly the US) tax incentives are available; these include tax credits and tax exemptions and can be awarded based on the costs of investment (the costs of the system and/or installation) or on the electricity generated by the system. Tax incentives may relate to a variety of different taxes including property tax, sales or goods and services tax (GST), personal tax, import tax and corporate tax. Tax incentives may be a set amount, a percentage of the total cost, or an amount per unit of energy production.

Loans

No or low-interest loans are available from a variety of institutions for the purchase of a PV system; these are sometimes referred to as green loans. The primary cost of a PV system is in the upfront purchase and loans programmes are designed to make owning a PV system possible for those without the capital. As well as being available from governments, loans are often available from banks and utilities. Loans are used in most parts of the world, including in developing countries.

Renewable portfolio standards and renewable energy certificates

Renewable portfolio standard (RPS) schemes are a common government policy intended to increase investment in renewable energy, particularly on a large scale. In an RPS scheme the government sets a target for renewable energy production, normally in the form of a certain percentage of electricity produced from renewable energy by a certain date, e.g. the Californian RPS targets 33 per cent renewable energy by 2020. In most schemes the large electricity utilities are required to comply and to prove they have done so they must surrender Renewable Energy Certificates (RECs), one REC typically representing 1MWh of electricity produced from renewable sources. REC is the term most commonly used in the US state-based RPS schemes, in the UK Renewable Obligation Certificates (ROCs) are used and in Australia Large-scale Generation Certificates (LGCs) and Small-scale Technology Certificates (STCs) are generated: which type of certificate is dependent on the size of the renewable energy system and the RPS scheme has separate quotas for LGCs and STCs. The utilities are required to surrender a sufficient number of RECs to cover the electricity from renewable sources required in that year. Under some US schemes, RECs can be traded on the open market and are known as Tradable Renewable Energy Certificates (TRECs), there may be a limit on the amount of TRECs a utility can surrender. Other schemes, i.e. Australia, make no such distinction and all LGCs and STCs can be traded on the open market. Some schemes also allow RECs to be banked for use in later years. A PV system will generate RECs and if the owner is not required to surrender RECs under the RPS scheme then they can sell their RECs to a utility. Sometimes these are referred to as Solar Renewable Energy Credits (SRECs). Small-scale PV systems will only generate a couple of RECs a year, but for a large-scale centralized PV system, selling RECs is a very important part of ensuring the system's profitability. REC prices fluctuate on the open market and the higher the REC price the more economically favourable the investment.

Box 13.5 Important features of RPS schemes

- Targets: RPS schemes are very widely used, but targets vary significantly as shown in Table 13.1.
- Banking: In some schemes excess RECs are bankable, i.e. a REC may be generated in 2008 but not surrendered until 2010. Schemes vary widely; banking may be allowed but limited to a certain number of years or unlimited, and in some a RECs deficit can even be carried forward and paid off in subsequent years.
- 'Bundled' and 'unbundled' RECs: 'Bundled' RECs are sold alongside the electricity, i.e. a system owner sells both the electricity from their system and the RECs generated to the utility. 'Unbundled' RECs can be sold separately, i.e. a utility has a surplus of RECs and so sells some to another utility; sometimes these are referred to as Tradable Renewable Energy Certificates (TRECs). Some RPS schemes have limits on the amount of TRECs that can be surrendered.
- Prices: In many RPS schemes the REC price is left to the market; however, some schemes incorporate upper and/or lower limits to guarantee some certainty to investors.
- Penalties: Penalties for non-compliance are generally per kWh; there is also often a limit to the amount of penalty a utility can incur within a year.
- Technology portfolios: Some RPS schemes have separate technology portfolios to encourage the different technologies: e.g. Japan aims to have installed 14GW of PV by 2020. Technology portfolios are fairly uncommon and so wind and hydro energy tend to dominate in RPS schemes as they are the least expensive renewable energy technologies.

Table 13.1 Renewable energy targets by location

Location	Target
Australia	20% by 2020
California, US	33% by 2020
China	15% by 2020
Connecticut, US	27% by 2020
Germany	18% by 2020
Spain	20% by 2020
Sweden	49% by 2020
Tokyo, Japan	20% by 2020
UK	15% by 2020
Washington, US	15% by 2020

Marketing

Installers of PV may need to 'sell' PV systems. Often the installer will only be approached once the decision to install a system has been made, but sometimes installers may need to convince potential clients of the benefits of PV. Knowledge of the economics of PV, as well as the environmental benefits will be useful here.

Grid-connected PV is the largest and most rapidly growing application of PV. This rapid increase is principally due to the financial incentives already discussed in this chapter. Financial incentives are a major marketing point for PV. The public often will not understand the limitations and extent to which these incentives will benefit them; it is the prerogative of the installer to explain the incentives to potential clients. The vast majority of systems will benefit from the FiTs, and installers should be very familiar with local legislation regarding these – it will be invaluable. Keeping up to date on any changes to legislation is also important.

The green image which PV imparts is also a major selling point – meaning clients may want their PV in a highly visible position at the expense of system performance. They may also want an attractive module, as opposed to an efficient one. Installers need to make the repercussions of these decisions absolutely clear to clients.

PV has many unique features among electricity-generating technologies that contribute to its popularity today:

- PV is reliable and low-maintenance; it contains no moving parts and the modules are robust, often coming with a 20–25 year guarantee.
- PV is good for the environment. Some people may suggest that a PV system cannot generate enough energy in its lifetime to equate to the amount of energy required in its production; this is not accurate. The energy required is retrieved in 2–7 years, depending on the system components, design, installation and its location. In addition this energy is clean and renewable, and will further reduce greenhouse gas emissions. PV modules are also recyclable.
- Grid-connected PV systems are easy and quick to install. PV's modular nature also makes it very easy to work with, installations can be of any size without any major manufacturing changes and modules can usually be added or removed during the lifetime of the installation.

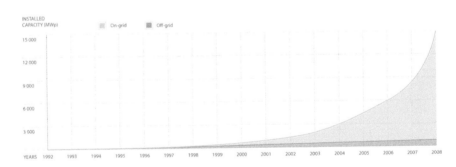

Figure 13.5 Graph showing the global cumulative installed PV capacity up to 2008. On-grid connections have increase dramatically over the last few years while off-grid has remained quite steady

Source: Global Sustainable Energy Solutions

- PV can represent a good investment: a system can add value to the building where it is installed and many of the financial incentives available such as FiTs are legislated for a certain period of time.
- PV is useful as a public demonstration of commitment to sustainability and reducing greenhouse gas emissions. This may be desirable for a company marketing itself as green because PV is an easy and highly visible way to make a statement and reduce a company's carbon footprint.

Despite these positive attributes there are still some strong barriers to PV technology, these include:

- High capital costs: the high capital cost of a system can be a major deterrent for investment or render a PV system unattainable for some people. Rebates and green loan programmes are attempting to address this.
- Lack of public knowledge and advertising: there are many myths surrounding PV such as it being too expensive to even consider putting on a home and that the amount of energy produced by the system is less than the energy consumed in its manufacture. The PV industry needs to be proactive about education to dispel these myths.
- Lack of industry: The grid-connect PV industry is still very much in its infancy in most parts of the world. There are a limited number of companies providing training and installation, and standards and regulations are still being developed.
- Lack of planning: As already discussed the amount of solar radiation a module receives is strongly affected by its orientation, but the installation is often governed by the orientation of the roof. When many towns were planned and built the roof orientation was not considered with respect to future PV installations.
- Utility regulations based on mains supply: Most electricity markets are unaccustomed to dealing with distributed generation, i.e. small-scale PV. Sometimes people seeking to install a small rooftop PV system have been required to complete the same paperwork as those planning to connect a large-scale coal-fired power station to the grid. Over time utilities are becoming more accustomed to dealing with distributed generation and streamlining the process of interconnection.
- Well-integrated fossil fuels: Well-established systems that favour fossil fuels are a key barrier to PV. Despite PV's numerous advantages, fossil fuels remain a more cost-effective option. Policies that put a price on the environmental damage of fossil fuels, such as emissions trading schemes and carbon taxes intend to drive up the price of fossil-fuel-generated electricity, making PV and other renewable energy sources more competitive.

Insurance

As with any major investment, PV arrays should be insured. As they are exposed to the open air, they are vulnerable to lightning, hail, excessive wind and vandalism. A well-designed system may be able to handle the first three, but unfortunately there is no guarantee against the fourth. Insurance costs need to be considered when determining the operating costs of the array. Local PV industry associations should be able to advise on insurance options.

Figure 13.6 Severe winds can cause damage to poorly anchored arrays

Source: Terry Cady

14

Case Studies

Case study A

Reference: RENAC Renewables Academy

Location: Berlin, Germany, latitude 53° north

Average annual solar irradiation: 1100kWh/m²/year on a plane tilted at 30°, orientated south. The average annual irradiation on a horizontal plane is approximately 950kWh/m²/year

Average daytime module temperature range: −10°C to 70°C

System

Three 1kWp arrays, each using different module types. RENAC is a training centre and has installed three different array types because they wanted to demonstrate and monitor the performance of three different technologies. Each array is connected to its own inverter (SMA Sunny Boy 1100). Arrays and inverters were matched using SMA design software. The system is monitored in the RENAC office several floors below the PV arrays. The system was installed in 2010.

PV Array 1

980Wp consisting of 7 × 140Wp Sunset TWIN thin film modules connected in series. The array's projected annual output is 850kWh.

PV array 2

1200Wp consisting of 6 × 200Wp Solon Blue polycrystalline silicon modules connected in series. The array's projected annual output is 1000kWh.

PV array 3

1150Wp consisting of 5 parallel strings, each string comprising 2 × 115Wp Inventux thin film silicon modules. This array also experiences partial shading from structures on the roof which could not be moved, especially in winter when the sun is low in the sky. This is the reason the thin film modules, which are less sensitive to shading, were chosen for this part of the roof rather than the crystalline ones. Projected annual output: 1000kWh/year – though the partial shading will reduce this to some degree.

Figure 14.1 Three inverters are used, one for each array

Source: Alberto Gallego

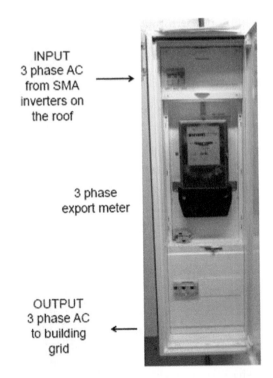

INPUT
3 phase AC
from SMA
inverters on
the roof

3 phase
export meter

OUTPUT
3 phase AC
to building
grid

Figure 14.2 A special three-phase export meter located in the basement of the building

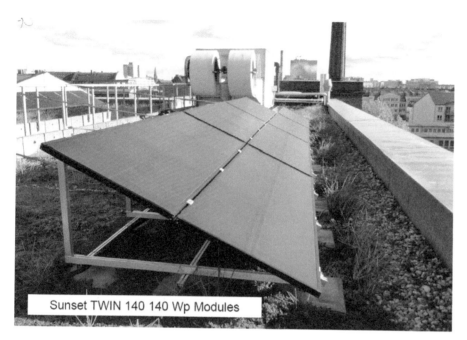

Sunset TWIN 140 140 Wp Modules

Figure 14.3 PV array 1

Source: Frank Jackson

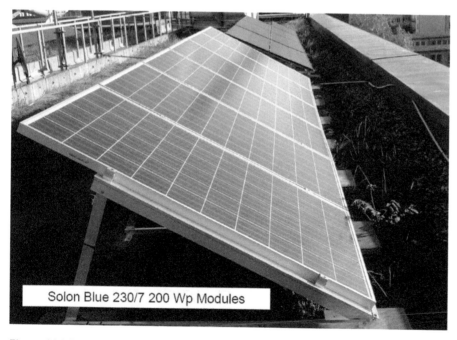

Solon Blue 230/7 200 Wp Modules

Figure 14.4 PV array 2

Source: Frank Jackson

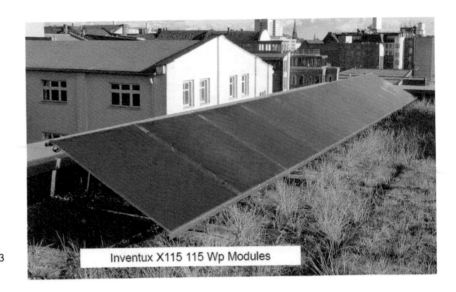

Inventux X115 115 Wp Modules

Figure 14.5 PV array 3

Source: Frank Jackson

Mounting structures

All the arrays are tilted at an angle of 30° using rack mounts on the flat roof. The structure is secured to the roof by concrete blocks (a technique known as ballast rack mounting).

Economics

Installed cost was €10,000 (excluding tax). RENAC is paid €0.39 for each kWh produced under the German renewable energy feed-in legislation.

Figure 14.6 Concrete blocks are used as ballast to weigh the system down to ensure it will not be affected by winds – the roof surface did not need to be punctured

Source: Frank Jackson

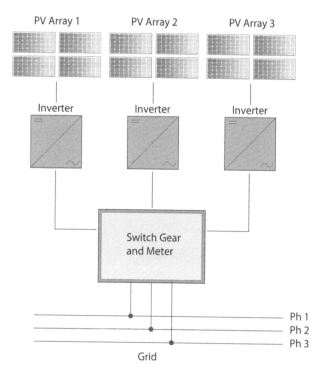

Figure 14.7 A block diagram of the RENAC system

Case study B

Reference: Residential property

Location: Welwyn, Hertfordshire, UK, latitude 52° north

Average annual solar irradiation: Approximately 970kWh/m²/year on a horizontal plane.

Average daytime module temperature range: 0°C–50°C

System

One 3.08kWp array linked to a 6kW inverter, the inverter is oversized to allow for future expansion of the PV array. The system has been sized to meet the electricity demands of the household and it is expected that annual electricity production from the PV system will meet annual electricity consumption. It was commissioned on 20 February 2010 and as of 20 January 2011 has generated 3113kWh of electricity with 2897Wh being metered at the distribution board.

PV array

The array comprises 14 Sharp ND220 polycrystalline silicon modules that are 13.4 per cent efficient. Array output varies significantly over the year: in the third week of June 2010 the system exported an average 17.2kWh/day, but it only exported 1.34kWh/day in the third week of December 2010 when the PV modules were covered by snow for part of the week.

Figure 14.8 The inverter is manufactured by Fronius and located indoors, under the stairwell within 1m of the distribution board

Source: Tony J. Almond, Planet Energy Solutions, Welwyn, UK

Figure 14.9 In order to meet the household energy demands 14 PV modules and 3 solar thermal collectors have been installed

Source: Tony J. Almond, Planet Energy Solutions, Welwyn, UK

Mounting structure

Standoff mounts have been used to allow some ventilation beneath the array. The roof faces south so the array could be installed at the tilt and orientation of the roof.

Economics

As of April 2010 the system owner receives £0.41/kWh under the UK Government's Clean Energy Cash Back scheme, a gross feed-in tariff (FiT). The

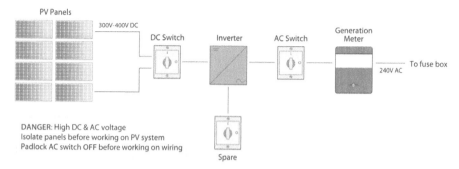

Figure 14.10 A wiring diagram of the PV system installed by Solar UK for Planet Energy Solutions, www.planetenergy.co.uk

Source: Tony J. Almond, Planet Energy Solutions, Welwyn, UK

income earned from this scheme (including the feed-in tariff and savings in electricity bills) is expected to amount to £1674/year, while it reduces CO_2 emissions by 1643kg/yr. From June 2011 the UK Government's Renewable Heat Incentive (RHI), a similar scheme to the FiT, will pay £0.18/kWh for the zero carbon heat generated by the solar thermal heating system, which at 20 January 2011 had generated 1774kWh of heat. The FiT and RHI incentive schemes mean that capital recovery will be less than 10 years for both systems.

Case study C

Reference: Commercial rooftop system

Location: San Diego, California, latitude 32° north

Average annual solar irradiation: Approximately 1400kWh/m²/year on a horizontal plane

Average ambient daytime module temperature range: 0°C–60°C

Figure 14.11 The greatest design challenge for this installation site was the limited available roof space. To overcome this the designer mounted modules facing south on the tile roof in addition to modules tilted south on the flat roof

Source: Global Sustainable Energy Solutions

System

This 57.12kWp system consists of 272 PV modules and 9 inverters. The designer of this system chose to use multiple small-scale inverters rather than a single commercial-scale inverter in order to reduce the effects of system downtime due to a potential inverter failure.

PV array

272 high-efficiency polycrystalline silicon Kyocera 210W modules were used. These modules have a +5/−0 per cent manufacturer's tolerance (see Chapter 9) guaranteeing their high performance.

Mounting structures

Two different mounting structures are used in this system. The modules on the tilted roof are mounted using standoffs, the small gap between the modules and the roof allows for cooling ventilation, which is very important in a warm climate such as California's.

Figure 14.12 The SMA inverters are mounted vertically on the south end of the building's parapet wall to limit their exposure to direct sunlight. These SMA inverters have an integrated DC disconnect switch (also known as a utility external disconnect switch) located just below the main body of the inverter as is common for PV systems in the US (see Chapter 5)

Source: Global Sustainable Energy Solutions

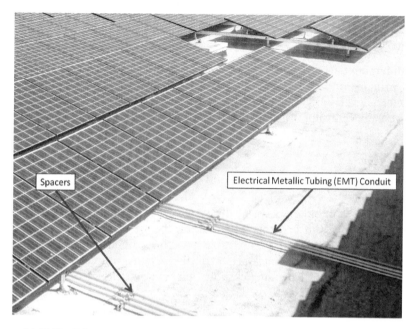

Figure 14.13 The DC conductors are housed inside a metallic conduit as per the guidelines in the National Electric Code (NEC). The 2008 NEC states that conduit in direct contact with a roof under full sunlight will experience an internal temperature of 33°C (60°F) above ambient while conduit that is mounted between 0.5 and 3.5 inches above the roof (as in this system) will experience an internal temperature of 22°C (40°F) above ambient. It is for this reason that the installers have mounted the conduit several inches above the surface of the roof

Source: Global Sustainable Energy Solutions

Figure 14.14 The mounting rails have been installed with periodic gaps between adjacent rails allowing the metal rails to thermally expand without damaging the mounting structure

Source: Global Sustainable Energy Solutions

Rack mounts are used for the module on the flat roof. This is a foam-sprayed rolled asphalt composition roof surface. In order to create well-sealed and watertight roof penetrations for the mounting system, the installers used an aluminium flashing around the standoff mount before applying a roof patch. The primary mounting system was designed around the ProSolar XD commercial rail. This high-strength rail is capable of supporting an 8-foot gap between adjacent support points, making it ideal for large systems.

Economics

The total system cost was US$305,000. On completion of the project the PV system received a rebate of US$70,968 under the California Solar Initiative (CSI). After rebate the net cost of the system was US$234,032 and the owner received a 30 per cent Federal Tax Credit on the net system cost (US$70,210). The final system cost was US$163,822, which is roughly 54 per cent of the initial system cost. The payback period of this system is estimated to be between five and seven years, depending on how quickly electricity rates increase in the local area.

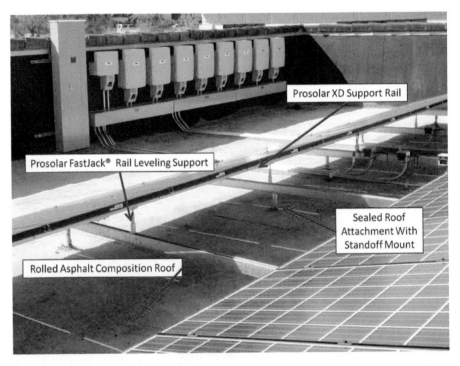

Figure 14.15 The Prosolar XD Support Rail holds the module while the ProSolar FastJack® rail levelling support is adjustable. This support height can be changed to modify the tilt angle of the PV module and ensure all the rails are on the same plane

Source: Global Sustainable Energy Solutions

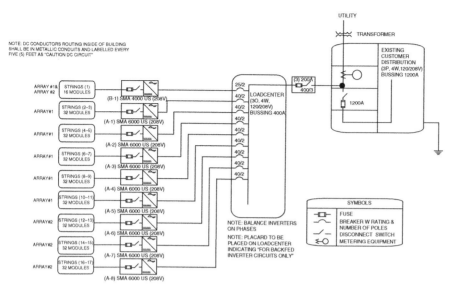

NOTE: DC CONDUCTORS ROUTING INSIDE OF BUILDING
SHALL BE IN METALLIC CONDUITS AND LABELLED EVERY
FIVE (5) FEET AS "CAUTION DC CIRCUIT"

Figure 14.16 Electrical diagram of the PV system

Case study D

Reference: Residential rooftop system

Location: Hinchinbrook, New South Wales, Australia, latitude −34° north or 34° south

Average annual solar irradiation: The average annual insolation on a horizontal plane is approximately 1650kWh/m²/year

Average daytime module temperature range: 0°C–70°C

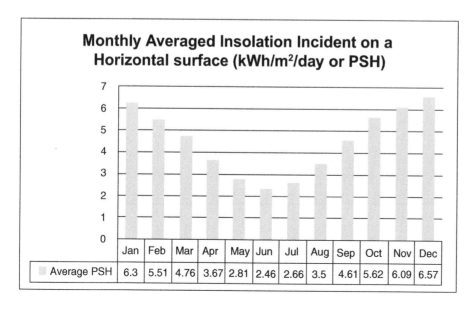

Monthly Averaged Insolation Incident on a Horizontal surface (kWh/m²/day or PSH)

	Jan	Feb	Mar	Apr	May	Jun	Jul	Aug	Sep	Oct	Nov	Dec
Average PSH	6.3	5.51	4.76	3.67	2.81	2.46	2.66	3.5	4.61	5.62	6.09	6.57

Figure 14.17 Average solar insolation in PSH or kWh/m²/day by month for Hinchinbrook, New South Wales

Source: NASA

System

A 2.22kWp system was installed in February 2010. The system uses a SMA Sunny Boy 2500 inverter.

PV array

The PV array consists of 12 Suntech 185Wp PV modules connected in series to form one string.

Mounting structures

The modules are mounted using standoffs and follow the tilt and orientation of the roof.

Economics

The capital cost of the system was AUD$19,680, on which the owners received a discount of AUD$960 from the installers. The owners received AUD$5100 under the Australian Government's Solar Credits Scheme so the net system cost was AUD$13,620. This system was also eligible for the New South Wales State Government's gross feed-in tariff which pays AUD$0.60/kWh (see Chapter 13 for further information on feed-in tariffs).

Figure 14.18 The PV array blends in with the profile of the roof and is aesthetically pleasing while allowing some cooling air flow beneath the modules

Source: Green Solar Group

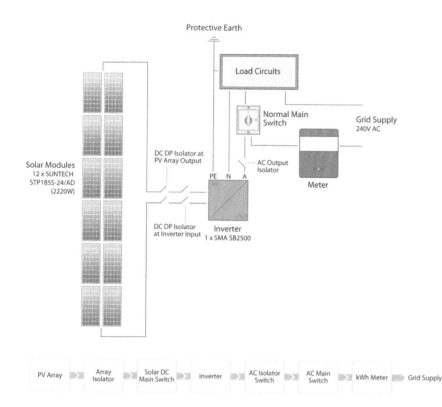

Figure 14.19 Diagram of the system installed by Green Solar Group

Case study E

Reference: Residential rooftop system

Location: Ramona, California, latitude 32° north

Average annual solar irradiation: The average irradiation on a horizontal plane is 5.8PSH/day, which is 1400kWh/m²/year.

Average daytime module temperature range: 0°C–60°C

System

This 7.65kWp system consists of 34 PV modules and 34 Enphase Energy M-190 micro-inverters. The system was designed with 8 modules on the east-facing roof at a 116° azimuth, 8 modules on the west-facing roof at a 210° azimuth, and 18 modules on the south-facing roof at a 164° azimuth. Micro-inverters were chosen because a PV system with a centralized inverter would have required at least 11 modules in a series string to face the same orientation and there was simply not enough roof space available for this. Using micro-inverters allowed the system designers to place any number of PV modules on each roof face. At this installation site, the south-facing roof also experienced some shading. If the system designers had chosen a centralized inverter, they

Figure 14.20 This flat tile residential rooftop presented some common design challenges for a typical PV system using a centralized inverter; there are several different roof orientations with suitable solar access (south, west and east). In addition there is not enough available space on each roof face to mount a complete series string, meaning that modules within the same series string would have to be mounted facing different orientations. Modules facing different directions would receive different amounts of solar radiation (see Chapters 7 and 9) and hence have different power outputs. As discussed in Chapter 4, it is not desirable to connect PV modules with very different outputs in series and the total power output of the system would suffer

Source: BMC Solar

would have had to install a specific minimum number of modules on each roof face. Given the limited suitable roof space, using the Enphase micro-inverters they were able to install modules in those roof areas least prone to shading.

PV array

34 high-efficiency back-contact monocrystalline silicon SunPower 225Wp modules were used. These modules have a module efficiency of 18.1 per cent and a manufacturer's tolerance of +5/−3 per cent (see Chapter 9).

Mounting structure

The mounting system used for this installation is SunFrame by Unirac. SunFrame is a low-profile, shared-rail mounting system that holds the module

Figure 14.21 In order to overcome the challenges presented by this residential roof, the system designer decided to forgo the traditional central inverter system architecture and install micro-inverters on each solar module. The micro-inverters are mounted on to the rails directly beneath the modules and convert the DC current from the PV module into AC current. Each individual micro-inverter has its own MPP tracking system so that the performance of each PV module is completely independent of its neighbours. The AC outputs are connected in parallel and then feed directly into the residential main service panel

Source: BMC Solar

Figure 14.22 In accordance with the National Electric Code (NEC), all metallic parts of the PV array which could potentially become exposed to live current are mechanically bonded to the earth (grounded/earthed) using either bare copper or green cables. This includes the PV module frames, mounting system and conduit as well as the individual micro-inverters as shown above

Source: BMC Solar

Figure 14.23 The mounting system achieves a good balance between aesthetics and performance. It is elevated slightly above the roof to allow for cooling ventilation but still retains a low profile and blends in with the roof face

Source: BMC Solar

Figure 14.24 The completed system takes advantage of the available roof space in the most efficient way possible. The sleek, all black PV modules and low-profile racking are considered aesthetically attractive

Source: BMC Solar

in place around the perimeter of the module's frame. Modules are supported in between SunFrame rails, not on top of them. Modules are slid into the SunFrame channel and are held firmly in place by the cap strip of the SunFrame mounting system.

Economics

The total system cost US$49,968 and upon completion of the project, the system owner received a rebate of US$4680 under the California Solar Initiative (CSI),

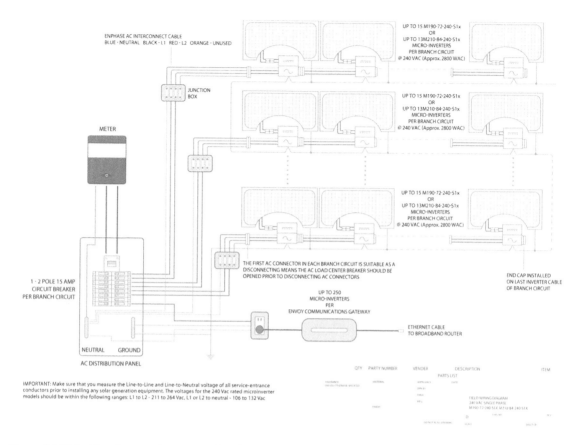

Figure 14.25 A wiring diagram of the system installed by BMC Solar

leaving a net system cost of US$45,288. They received a 30 per cent Federal Tax Credit on the net system cost, which equated to US$13,586. Following these financial incentives the net system cost was US$31,702, which is approximately 63 per cent of the initial cost. The payback period is estimated to be 8–10 years.

Case study F

Reference: Residential rooftop system

Location: Cumnock, New South Wales, Australia, latitude −33° north or 33° south

Average annual solar irradiation: The average irradiation on a horizontal plane is approximately 5.11PSH/day or 1850kWh/m²/year

Average daytime module temperature range: 0°C–70°C

System

This is a 9.84kWp system installed on a property in rural Australia. The system uses 2 Aurora Powerone PVI-500 outdoor inverters each connected to 24 modules (there are 48 modules in total). Each inverter contains two Maximum Power Point Trackers. The system uses a digital meter located far from the array and inverter (it is common in large rural properties for the switchboard/distribution board and meter to be located at the front gate, which is some distance from the installation) which requires a 65m-long underground cable run.

Figure 14.26 The system included an underground cable run, 16mm² SDI cable was used

Source: Green Solar Group

Figure 14.27 The digital meter

Source: Green Solar Group

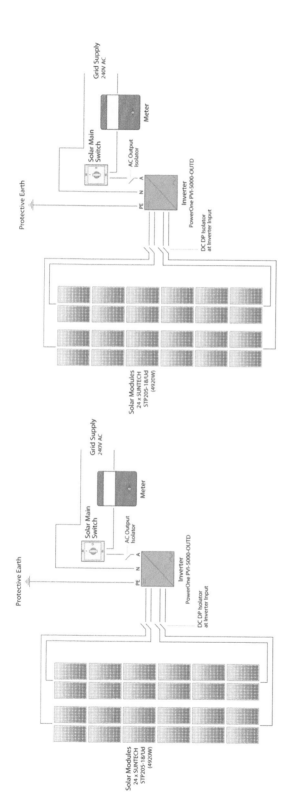

Figure 14.28 A diagram of the PV systems showing the two separate arrays installed by Green Solar Group

Figure 14.29 In this photograph the metal rails that make up the mounting structure are clearly visible

Source: Green Solar Group

PV array

There are two identical 4.92kWp PV arrays connected to two separate inverters. Each PV array consists of 24 × 205Wp polycrystalline Suntech modules connected in 2 series strings (each of 12 modules), which are connected in parallel.

Mounting structure

The modules were mounted using standoff mounts. Fortunately the roof faces due north and is tilted at 30°, which is ideal for the PV array. It does, however, experience some shading (~5%).

Economics

The capital cost of the system was AUD$66,075, including a cost of AUD$325 for the underground cable run. The system received AUD$13,120 under the Australian Government's Solar Credits Scheme so the net system cost was AUD$52,925. This system was also eligible for the New South Wales State Government's gross feed-in tariff (FiT), which pays AUD$0.60/kWh, and is expected to receive AUD$8,560 per annum in FiT payments. The system is expected to pay itself off in six to seven years and save an estimated 16,120kg CO_2 per annum.

Case study G

Reference: Stanton House

Location: Pembury, Kent, UK, latitude 51° north

Average annual solar irradiation: Approximately 1150kWh/m²/year on a plane tilted at 30° orientated south. The average annual irradiation on a horizontal plane is approximately 1000kWh/m²/year

Average daytime module temperature range: 0°C–50°C

Figure 14.30 The PV array is ground-mounted in the owner's garden

Source: Paul Barwell

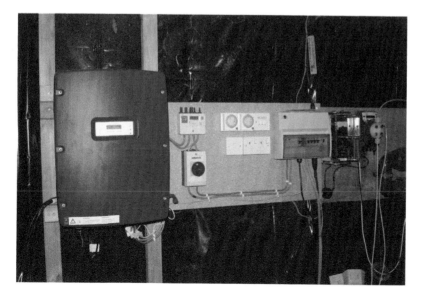

Figure 14.31 The inverter is located in the pool shed behind the PV array; the inverter is connected to a generation meter which measures the PV array's power output, then to a lockable double pole isolating switch and then directly into one of the miniature circuit breakers located on a distribution board in the shed

Source: Paul Barwell

System

This is a 5.04kWp system installed on a large residential property in the UK. The house has an east/west facing roof so the installer (who was also the system owner) decided to mount the modules on the ground, which meant they could face due south at a tilt of 30°. Aesthetics were important to the system owner so the modules were mounted at the edge of the garden, which has since been landscaped so that they cannot be seen from the house or the road. The modules are connected to an SMA Mini Central Inverter which is located in the pool shed behind the array. The system also includes an SMA Sunny Beam display and wireless piggyback monitoring system.

PV array

The array comprises 24 Sanyo heterojunction with intrinsic thin layer (HIT) modules (see Chapter 3), which were chosen because the system owner valued reliability over expense and Sanyo have an excellent reputation. These are installed in two rows of 12 modules; however, electrically they are connected in three strings of 8 modules per string. Each series string has 4 modules from the front row and 4 modules from the back. The SMA website (www.sma.de) was used to size the array and match it to the inverter. The PV array does occasionally experience shading; the front row of modules shades the back in the middle of winter and surrounding trees shade the PV array at the beginning and end of the day in early and late summer. In summer (May) the array is often able to generate around 34kWh/day, but in winter array output decreases to about 10 per cent of that experienced in summer.

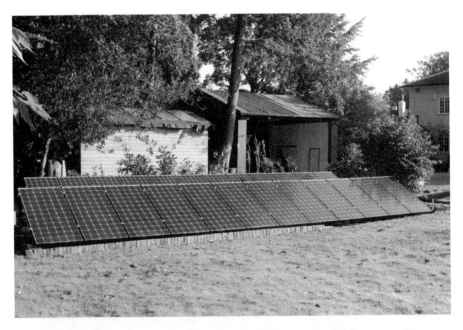

Figure 14.32 The PV array is surrounded by trees and occasionally experiences shading

Source: Paul Barwell

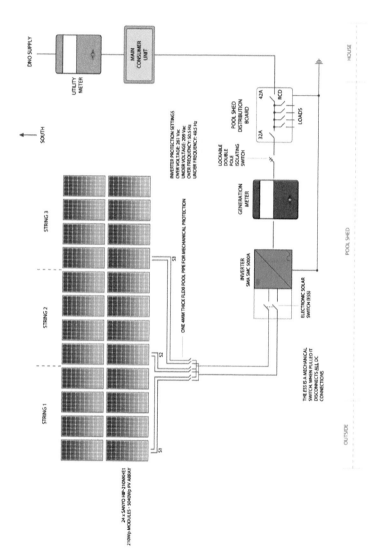

Figure 14.33 A block diagram of the PV array installed by Paul Barwell

Mounting structure

The modules are ground-mounted; the installer used 16 preservative-treated (Tanalised®) oak railway sleepers as a base, laid across flat ground on a weedproof sheet that was later covered with gravel for aesthetic reasons. The

Figure 14.34
Construction of the mounting structure, the wooden A-frames and Unistrut strip (along the top) can be seen. Another strip of Unistrut can be seen on the grass; it will be attached along the bottom of the A-frames

Source: Paul Barwell

Figure 14.35 The mounting structure viewed from the side

Source: Paul Barwell

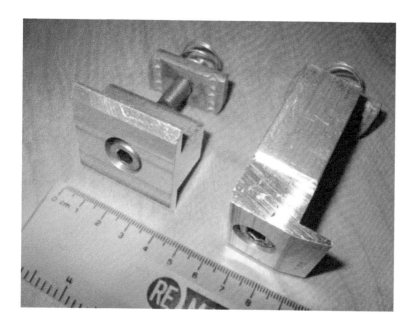

Figure 14.36 Conergy clamps are used to secure the modules to the Unistrut

Source: Paul Barwell

railway sleepers were chosen for their durability; preservative pressure-injected wood is expected to last at least 25 years. Sixteen Tanalised® wood A-frames were constructed and screwed onto the sleepers. Two strips of Unistrut hot-dipped galvanized steel were joined to the top and bottom of the A-frames and the modules were fixed to the Unistrut using Conergy clamps.

Economics

The total system cost was £20,900, which included £15,700 for the PV modules and £2,200 for the inverter. The system feeds electricity into the grid and earns £0.16/kWh generated under a gross feed-in tariff paid by the system owner's utility (Ecotricity Ltd). In its first year of operation the PV system generated 4200kWh, which equates to an earning of £672. The payback period is estimated to be 17 years; however, if the system were installed now (January 2011) it would receive a gross feed-in tariff of £0.41/kWh under the UK Government's Clean Energy Cash Back scheme (as in case study B), which would reduce the payback period to approximately 7 years.

Figure 14.37 The PV array is cleverly hidden in the garden so that it does not affect the appearance of this beautiful property

Source: Paul Barwell

15

Grid-connected PV and Solar Energy Resources

PV and renewable energy news

Home Power Magazine

This bimonthly magazine is available in the US and provides information regarding installation practices. It is targeted at all levels, from beginner to professional; www.homepower.com.

Sun and Wind Energy Magazine

This magazine includes good general information on current developments in renewable energy; www.sunwindenergy.com.

PVTECH

An industry-sponsored website featuring interesting articles on the latest developments in PV technology; www.pv-tech.org.

Semiconductor Today

This website features news on the latest developments in the electronics industry including photovoltaics; www.semiconductor-today.com.

Greentech Media

This website features news and discussion on current and emerging green technologies including photovoltaics; www.greentechmedia.com.

Renewable Energy World

A magazine focused on a wide range of topics relevant to renewable energy, including company and product information, job advertisements, blogs, podcasts and news on global developments within the industry. Both print and PDF versions are available; www.renewableennergyworld.com.

PV Magazine

A monthly trade publication for the international PV community, it has independent, technology-focused reporting, concentrating on the latest PV news, topical technological trends and worldwide market developments; www. pv-magazine.com.

Further reading on PV technologies

Buried contact solar cells

Buried contact solar cells were developed at the University of New South Wales (Australia) and further information can be found on their website; www.pv. unsw.edu.au/info-about/research-5.

National Renewable Energy Laboratory (NREL)

NREL is the primary laboratory for renewable energy and energy efficiency research and development in the US. The website offers information regarding current and emerging renewable energy technologies; aimed at a wide audience, it is an excellent site for anyone interested in learning more; www.nrel.gov.

US Department of Energy Efficiency and Renewable Energy

An excellent resource for those seeking basic information regarding renewable energy and energy efficiency measures such as low-energy building design. This website also includes information regarding government policy and incentives for renewable energy; www.eere.energy.gov.

The books listed in Table 15.2 are only a few of those published by Earthscan on the topic of solar energy. They have provided invaluable research material in the writing of this book.

Table 15.1 Resources for emerging PV technologies

Technology	Website
Dye-sensitized solar cells	Dyesol (manufacturer); www.dyesol.com
Sliver cells	Transform Solar (manufacturer); www.transformsolar.com/tech_sliver.php and Australian National University (research); http://cses.anu.edu.au/docs/Slivers.pdf
HIT PV cells	Sanyo (manufactuer); http://Solar.sanyo.com/hit.html
III-V semiconductors	Spectrolab (manufacturer); www.spectrolab.com and Emcore (manufacturer); www.emcore.com

Table 15.2 Earthscan publications

Title	Description
Photovoltaics for Professionals	Targeted at professionals in the PV industry (electricians, builders, architects and engineers), this book gives an in-depth description of the design and installation processes, covering mainly grid-tied PV systems
Applied Photovoltaics	Essential reference material for any engineer working in the field of photovoltaics, this book covers semiconductors and p-n junctions, cell properties and module fabrication, in addition to stand-alone photovoltaic systems, grid-connected photovoltaic systems and photovoltaic water pumping system components and design
Planning and Installing Photovoltaic Systems	A thorough guide to every aspect of grid-connected photovoltaic systems, *Planning and Installing Photovoltaic Systems* is an invaluable resource for installers, architects and engineers working in the field of photovoltaics
Solar Domestic Water Heating (Expert Series)	A fully illustrated and easy-to-follow guide introducing all aspects of solar domestic water heating systems, including their operation and installation
Stand-alone Solar Electric Systems (Expert Series)	A practical guide to the design and installation of stand-alone PV systems, written for electric technicians and designers, development workers and the DIY enthusiast

PV in major English-speaking nations

International

- International Solar Energy Society; www.ises.org
- The Renewable Energy and Energy Efficiency Partnership; www.reeep.org

Australia

- Australian PV Association; www.apva.org.au
- Australian Solar Energy Society; www.auses.org.au
- Clean Energy Council (Australia); www.cleanenergycouncil.org.au
- Department of Climate Change and Energy Efficiency; www.climatechange. gov.au
- Global Sustainable Energy Solutions (GSES), training provider; www.gses. com.au

Canada

- Canadian Solar Industries Association; www.cansia.ca

Republic of Ireland

- Department of Communications, Energy and Natural Resources; www.dcenr.gov.ie/Energy/Sustainable+and+Renewable+Energy+Division

South Africa

- National Energy Regulator of Southern Africa; www.nersa.org.za
- Sustainable Energy Society of Southern Africa; www.sessa.org.za

UK

- Centre for Alternative Technology (Wales), training provider; www.cat.org.uk
- Department of Energy and Climate Change; www.decc.gov.uk
- Green Dragon Energy, training provider; www.greendragonenergy.co.uk
- UK Solar Energy Society; www.uk-ises.org

US

- American Solar Energy Society; www.ases.org
- Contractors License Reference Site; http://contractors-license.org
- Council of American Building Officials and Code Administrators; www.bocai.org
- Database of State Incentives for Renewable Energy and Energy Efficiency, information on federal, state, local and utility incentives for renewable energy; www.dsireusa.org
- Florida Solar Energy Council; www.fsec.ucf.edu/en
- Illinois Solar Association; www.illinoissolar.org
- Interstate Renewable Energy Council (IREC); www.irecusa.org
- Midwest Renewable Energy Association, training provider; www.the-mrea.org
- North American Board of Certified Energy Practitioners (NABCEP); www.nabcep.org
- Oregon Solar Association; www.oseia.org
- Solar Energy International, training provider; www.solarenergy.org
- Solar Living Institute, training provider; www.solarliving.org

Standards and certification for PV modules and system components

- Underwriters' Laboratory (UL); http://ul.com/global/eng/pages/offerings/industries/energy/renewable/photovoltaics
- TÜV; www.tuv-pv-cert.de/en/certificates-of-pv-modules.html
- PV Resources; www.pvresources.com/en/standards.php
- European Solar Test Installation (ESTI); http://re.jnc.eu.europa.en/esti/index_en.htm
- International Electrotechnical Commission (IEC); www.iec.ch/renewables/standardization.htm
- Solar Rating and Certification Corporation; www.solar-rating.org

Installation codes and guidelines

International Electrotechnical Commission (IEC) standards

- IEC 61215 Crystalline silicon terrestrial photovoltaic (PV) modules – Design qualification and type approval
- IEC 61345 UV test for photovoltaic (PV) modules
- IEC 61646 Thin-film terrestrial photovoltaic (PV) modules – Design qualification and type approval
- IEC 61701 Salt mist corrosion testing of photovoltaic (PV) modules
- IEC 61730–1 Photovoltaic (PV) module safety qualification Part 1: Requirements for construction
- IEC 61730–2 Photovoltaic (PV) module safety qualification Part 2: Requirements for testing
- IEC 61829 Crystalline silicon photovoltaic (PV) array – On-site measurement of I-V characteristics
- IEC 62108 Concentrator photovoltaic (CPV) modules and assemblies – Design qualification and type approval
- IEC 62446 Grid-connected PV systems – Minimum requirements for system documentation, commissioning tests and inspection

Australia and New Zealand

- Australian/New Zealand Standard 4777 – Grid Connection of Energy Systems via Inverters
- Australian/New Zealand Standard 5033 – Installation of Photovoltaic Arrays
- *Grid-Connected PV Systems Design and Installation*, 7th Australian edn, Stapleton, G., Garrett, S., Neill, S. and McLean, B. (2010), Global Sustainable Energy Solutions, Sydney

Singapore

- Code of Practice for Electrical Installations, CP 5: 1998 – *Electrical Installations*
- Singapore Standard 555:2010 – *Code of Practice for Protection against Lightning*
- Singapore Standard 371:1998 – *Specifications of Uninterruptible Power Supplies*

UK

- British Engineering Standard Committee; www.bsigroup.com
- British Standard 476 *Fire tests on building materials and structures*
- British Standard 3535 *Specification for safety isolating transformers for industrial and domestic purposes*
- British Standard 3858 *Specification for binding and identification sleeves for use on electric cables and wires*

- British Standard 7671 *Requirements for electrical installations, IEE Wiring Regulations*
- British Standard 60947 *Specification for low-voltage switchgear and controlgear*
- Engineering Recommendation G83/1 (2003) – Recommendations for the connection of small-scale embedded generators (up to 16A per phase) in parallel with public low-voltage distribution networks
- IEE Guidance Note 7 to BS 7671 – Special Locations, Section 12 Solar Photovoltaic (PV) Power Supply Systems (2003)
- *Photovoltaics in Buildings: Guide to the Installation of PV Systems* 2nd edn, Department for Enterprise UK, DTI/Pub URN 02/788; www.bre.co.uk/filelibrary/pdf/rpts/Guide_to_the_installation_of_PV_systems_2nd_Edition.pdf

US

- *A Guide to Photovoltaic System Design and Installation*, Bill Brooks, California Energy Commission Consultant Report 500–01–020, June 2001; www.energy.ca.gov/reports/2001-09-04_500-01-020.PDF
- *Connecting to the Grid: A Guide to Distributed Generation Interconnection Issues*, Interstate Renewable Energy Council (IREC); irecusa.org/fileadmin/user_upload/ConnectDocs/Connecting_to_the_Grid_Guide_6th_edition-1.pdf
- North American Board of Certified Energy Practitioners (NABCEP), Solar PV References; www.nabcep.org/resources
- *PV Codes and Standards* Website, John Wiles, Southwest Technology Development; www.nmsu.edu/~tdi/Photovoltaics/Codes-Stds/Codes-Stds.html
- *The 2011 National Electric Code*

Wind loading codes and guidelines

International

- Draft International Standard (local standards take precedence): ISO/DIS 4354–2007
- EN 1991-Part 1–4: 2009, Eurocode 1: Actions on structures: Part 1–4: Wind actions

Australia and New Zealand

- Australian/New Zealand Standard 1170.2 – Wind Loading

Singapore

- Singapore National Annex to Singapore Standard EN 1991–1-4:2009

South Africa

- South African National Standard 10160 – Basis of Structural Design and Actions for Buildings and Industrial Structures

UK

- BRE – Digest 489 Wind loads on roof-based photovoltaic systems
- BRE – Digest 495 Mechanical installation of roof-mounted photovoltaic systems
- British Standard 6399 *Loading for buildings: Code of practice*

US

- American Society of Civil Engineers (ASCE) Standard 7–05: Minimum Design Loads for Buildings and Other Structures

Solar resource data and simulation software

- PVGIS EU Joint Research Centre provides meteorological data for Europe and Africa and an online simulation tool; http://re.jrc.ec.europa.eu/pvgis
- PV*SOL: Design and sizing software commonly used in the PV industry, a demo version of PV*SOL can be downloaded free; www.valentin.de
- PVsyst: This programme can be used for the design and simulation of both grid-connected and stand-alone systems, it is incredibly useful for sizing (see Chapter 9). A demo version can be downloaded free; www.pvsyst.com
- NASA Surface Meteorology and Solar Energy Data Set, solar radiation data (on a horizontal plane) for all locations; http://eosweb.larc.nasa.gov/sse
- Solar Radiation Data Manual for Flat-Plate and Concentrating Collectors, National Renewable Energy Laboratory (NREL, US); rredc.nrel.gov/solar/pubs/redbook
- Sunny Design: Sizing software for use with SMA inverters, free to download; www.sma.de

16
Glossary

AC (electricity): Alternating current. An electric current in which the direction of current flow alternates at frequent intervals. Mains/grid electricity is AC; in Europe the direction of mains current alternates 100 times per second (at a frequency of 50Hz), while in Australia and the US it alternates 120 times per second (at a frequency of 60Hz).

Air mass: Air mass relates to the distance that solar radiation must travel through the atmosphere. An air mass of 1 means that the sun is directly overhead and must travel through a thickness of 1 atmosphere to reach a point on the Earth's surface. This value varies throughout the day for any given location and is calculated using trigonometry and the solar *altitude angle*.

Albedo: The amount of solar radiation reflected from the Earth's surface at a given location.

Altitude angle (solar): The angle between the sun's position in the sky and the horizon at a given time.

Ambient temperature: The temperature of the surrounding environment.

Ampere (amp or A): The unit of measurement for electrical current.

Amorphous (silicon): A non-crystalline substance. Amorphous solids lack order and structure in their molecular composition. Glass is as example of an amorphous solid; so is amorphous silicon, which is used in some thin film PV cells.

Azimuth: The angle between north and the point on the compass where the sun is positioned. The azimuth is measured clockwise from true north along the horizontal plane (along the ground).

Balance of system (BoS) equipment: Referring to the components of a PV system excluding the PV modules and inverter. BoS equipment includes the mounting structure, cabling, disconnects/isolators, module junction boxes, PV combiner boxes, grounding/earthing equipment and meters.

Bypass diode: A solid state electrical component able to pass current in one direction only and which will allow current to bypass a shaded or damaged cell so that it does not hinder the output of the other cells.

Circuit breaker: A device designed to automatically interrupt the flow of current when an electrical fault occurs. This device usually includes a method of extinguishing an electrical arc if it forms.

Concentrated Solar Power (CSP): This is a large-scale form of *solar thermal technology* that creates electricity by using mirrors to concentrate the sun's rays. In most applications the sun's rays are then used to heat water and create steam to drive a steam turbine and generate electricity.

Current (I): The flow of electrons around a circuit. Conventional current flows from positive to negative. The SI unit of current is the ampere (A).

DC (electricity): Direct current. An electric current that does not vary periodically in magnitude, but rather flows steadily in one direction. PV cells produce DC.

Deciduous: Refers to trees that lose their leaves during winter.

Direct radiation: Solar radiation that arrives at the Earth's surface directly from the sun.

Diffuse radiation: Solar radiation that is scattered or absorbed by clouds and gases within the atmosphere and then re-emitted before reaching the Earth's surface.

Dye-sensitized solar cells: An emerging technology that uses coloured dyes and titanium dioxide to produce electricity.

Efficiency – PV Cells: A measurement of the power output (electricity) compared to the power input (solar radiation), efficiency is often used when comparing PV cells.

Electromagnetic radiation (EMR): Energy that travels through space as a wave. Sunlight is an example of EMR.

Fuse: A device that protects electrical systems from damage by cutting off the electricity supply in the event of a short circuit or current overload.

Grid-connected system: A PV system that can export power directly to an electrified grid. Typically there is no storage (i.e. batteries) in a grid-connected system.

Ingot (silicon): A block of pure silicon from which PV cells are cut. The ingot may be either *multicrystalline* or a single large crystal (*monocrystalline*), depending on the casting method.

Insolation: The amount of solar radiation incident on a surface over a day, measured in peak sun hours (PSH) or kWh/m²/day.

Inverter: A device that converts DC electrical power into AC, the inverter is an essential component in grid-connected photovoltaic systems.

Irradiance: The amount of solar radiation incident on a surface at any one time, measured in W/m² or kW/m².

Irradiation: The total quantity of solar energy per unit area received over a given period, e.g. daily, monthly or annually. Irradiation is typically measured in W/m²/time period. Annual irradiation is commonly given for a site and is measured in kWh/m²/year. The term insolation refers to irradiation measured over a day. Irradiation is a cumulative measurement rather than an instantaneous measurement; it is the sum of irradiance over a period of time.

I-V curve: A graph used to plot the output characteristics of a PV cell. The plot shows voltage versus current and can be used to determine power output and efficiency.

kWh: Kilowatt-hour, a common unit for measuring energy.

Load: Refers to the amount of power being consumed at any given time and/or devices that consume power.

Low voltage (LV): The IEC defines low voltage as any voltage range between 50 and 1000V AC or 120 and 1500V ripple-free DC. See *volt (V)*.

Magnetic declination/variation: The difference between true geographical north and magnetic north (as on a compass), this value varies significantly depending on location.

Maximum power point (MPP): The point on the I-V curve at which the most power is extracted from the photovoltaic module.

Maximum power point tracker (MPPT): A device which modifies the load on a photovoltaic array to allow the system to operate at its *maximum power point*.

Monocrystalline: Refers to a silicon wafer made up of a single crystal. This is typically produced from a silicon 'seed' crystal which is placed in a crucible of molten silicon and drawn out slowly while rotating.

Multicrystalline: Also known as *polycrystalline*, multicrystalline PV cells are made from silicon wafers composed of many crystals. Typically these are produced by block casting molten silicon, which allows many small silicon crystals to form.

Nominal operating cell temperature (NOCT): The temperature at which the cells in a PV module will operate under the following conditions: $800W/m^2$ irradiance, 20°C ambient temperature and a wind speed of 1m/s.

Nuclear fusion: The reaction taking place within the sun, where atoms combine or fuse to produce an atom of a different element, i.e. two hydrogen atoms combine to form a helium atom.

Off-grid system: Also known as *stand-alone power system*. Off-grid systems are not connected to the mains power supply, and typically store power in batteries for later use.

Ohm (Ω): The SI unit for electrical resistance.

Open-circuit voltage (V_{oc}): The voltage produced across a PV cell or module in direct sunlight when no current is flowing, i.e. the maximum voltage a PV cell may produce. V_{oc} is given on manufacturers' data sheets and is measured in *volts (V)*.

Passive solar design: Refers to the design of buildings to utilize the sun's energy effectively, typically to increase thermal comfort and reduce heating and cooling loads.

Peak power: The maximum amount of a power a PV cell, module or array is expected to produce under *standard test conditions*. Peak power is typically given in watts-peak (Wp), kilowatts-peak (kWp) or megawatts-peak (MWp).

Peak sun hours (PSH): A common unit of measurement for *insolation* or daily *irradiation*. The number of PSH for the day is the number of hours during which power at the rate of $1000W/m^2$ is delivered; therefore it gives an equivalent amount of energy to the total energy for that day, i.e. it is the daily irradiation in $W/m^2/day$ divided by 1000.

Photoelectric effect: The process by which electrons are emitted from a substance due to its absorption of electromagnetic radiation. See *electromagnetic radiation*.

Photosynthesis: The process by which sunlight, carbon dioxide and water are converted to carbohydrates. Photosynthesis is the means by which plants survive.

Photovoltaic (PV): Refers to devices that produce electricity from direct sunlight.

Photovoltaic array: Solar/PV modules are physically and electrically connected to form an array.

Photovoltaic cell: A single semiconductor device that generates an electric current when sunlight is incident on its surface; also known as a solar cell.

Photovoltaic module: Photovoltaic cells are physically and electrically connected to form a module. The cells are typically held together by a frame and encapsulated in a protective substance (often glass).

Polycrystalline: See *multicrystalline*.

Potential difference (electricity): The difference in potential energy between two points. If these points are connected by a conductor, current will flow. The SI unit of electrical potential difference is the volt (V).

Resistance: The measure of a material's capacity to oppose the flow of electric current. The SI unit of resistance is the ohm (Ω).

Semiconductors: A type of material from which PV cells and many electrical components are manufactured. Semiconductors have properties of both insulators and conductors: silicon and germanium are examples.

Short-circuit current (I_{sc}): The current flowing through a PV cell under short-circuit conditions, i.e. when there is no load or resistance. This is the maximum current a PV cell may produce and is measured in amps (A).

Silicon: A semiconductor material commonly used to make PV cells.

Solar concentrators: Devices which increase the intensity of light hitting a solar cell; they are typically lenses or reflective troughs. They can be used to increase the power output of a PV cell by increasing the solar energy incident on the cell.

Solar constant: This is the peak irradiance value of solar radiation arriving at the Earth's atmosphere ($1367W/m^2$).

Solar electric system: Systems that convert the sun's energy directly into electricity. Photovoltaic (PV) systems are the main technology in this category.

Solar thermal technology: Systems that capture the sun's heat, such as solar hot water or *concentrated solar power*.

Stand-alone power system: See *off-grid system*.

Standard test conditions (STC): A set of standard conditions under which PV cells are tested so that they can be compared. These conditions are $1000W/m^2$ irradiance, air mass of 1.5 and a cell temperature of $25°C$.

SI unit: Abbreviated form of Système International d'unités (International System of Units), a series of units that allows for easy engineering calculations. This group includes the watt, joule and metre, as well as most other units used in this publication.

Thermal mass: Describes a building's resistance to fluctuations in external temperature. A material with high thermal mass is described as one which absorbs and re-emits thermal energy slowly, e.g. concrete. A material with low thermal mass is one which absorbs and re-emits thermal energy quickly, e.g. paper.

Thin film solar cells: Solar cells made from materials that are suitable for deposition on large area materials. They are known as 'thin film' because the layer of semiconductor material is significantly thinner than the regular wafer used in mono- or multi-crystalline solar cells.

Tracking systems: Mounting systems that include mechanical devices which can alter the orientation axis and/or tilt axis of the PV array in order to optimize the exposure of the array to the sun and capture more solar radiation.

Volt (V): The unit of measurement for voltage or electrical potential difference. See *potential difference (electricity)*.

Voltage drop: The loss of voltage (and therefore power) due mainly to the resistance of cables.

Watt (W): The SI unit of measurement for power. It is common for PV systems to be described in kilowatts (kW), equal to 1000W, or megawatts (MW), equal to 1,000,000W.

Watt-hour (Wh): A unit of measurement of energy. An appliance using 1 watt of power for 1 hour will use 1Wh of energy. The same device operating for 6 minutes will use 0.1Wh. Kilowatt-hours (kWh), equal to 1000Wh, are also commonly used.

Watt-peak (Wp): The output of a PV cell under optimum conditions, usually standard test conditions. Kilowatt-peak (kWp) and megawatt-peak (MWp) are commonly used to describe the output of a PV array. see *standard test conditions (STC)*.

Index

Printed and bound by CPI Group (UK) Ltd, Croydon, CR0 4YY

01/11/2024

01782604-0005